The Institute of Biology's
Studies in Biology no. 165

Seed Physiology

John A. Bryant
Professor of Biological Sciences
University of Exeter

Edward Arnold

First published in Great Britain 1985 by
Edward Arnold (Publishers) Ltd,
41 Bedford Square, London WC1B 3DQ

Edward Arnold (Australia) Pty Ltd,
80 Waverley Road, Caulfield East, Victoria 3145, Australia

Edward Arnold, 3 East Read Street, Baltimore,
Maryland 21202, USA

British Library Cataloguing in Publication Data

Bryant, John A.
Seed physiology.—(The Institute of Biology's studies in biology,
ISSN 0537-9024; no. 165)
1. Seeds—Physiology
I. Title II. Series
582′.0467 QK661

ISBN 0-7131-2898-4

Text set in 9½/11 English Times Compugraphic
by Colset Pte. Ltd, Singapore
Printed and bound in Great Britain at
The Camelot Press Ltd, Southampton

General Preface to the Series

Because it is no longer possible for one textbook to cover the whole field of biology while remaining sufficiently up to date, the Institute of Biology proposed this series so that teachers and students can learn about significant developments. The enthusiastic acceptance of 'Studies in Biology' shows that the books are providing authoritative views of biological topics.

The features of the series include the attention given to methods, the selected list of books for further reading and, wherever possible, suggestions for practical work.

Readers' comments will be welcomed by the Institute.

1985

Institute of Biology
20 Queensberry Place
London SW7 2DZ

Preface

In this little book I have attempted to present, for general readership, current ideas about the physiology of seeds, their development on the parent plant, dormancy mechanisms and their adaptive relevance, the process of germination, and finally the importance of seed physiology to mankind. I have drawn on a wide variety of sources, including original research articles, review articles, books and in some instances, as yet unpublished data. I trust that my distillation of these sources will provide a stimulating introduction to a topic which I personally find fascinating.

In writing any book, an author relies not just on himself, but also on others. In this context I wish to thank the many people who have, knowingly or unknowingly helped me in my task: my former academic colleagues at University College, Cardiff, particularly Dr John Etherington; colleagues in other Universities and Research Laboratories, particularly the late Professor James Sutcliffe, Professor Mike Black and Drs Cliff Bray, Paul Brocklehurst, John Chapman, Don Grierson, and Neville Pinfield; past and present members of my own research group, particularly Drs Peter Coolbear, Sally Greenway, Joseph Grey, Robert Slater, Geoffrey Strangeway and Sheila Thompson.

It is also a pleasure to thank Mrs Hilary Webb who translated my heiroglyphics into a readable typescript, and the staff of Edward Arnold for their encouragement and patience.

Exeter, 1985

J.A.B.

Contents

Introduction

'. . . a grain of wheat remains a solitary grain unless it falls into the ground and dies; but if it dies it bears a rich harvest.'

St. John's gospel, Ch. 12 v. 24, New English Bible.

The opening quotation, taken out of its context as an illustration of a spiritual theme, indicates very clearly that in biblical times, seeds of crop plants were both familiar and mysterious. These elements of familiarity and mystery remain today as a fair summary of the knowledge that many people have of seeds.

Seeds are familiar because many people are introduced to seeds at an early age. Even in inner city areas, far from the countryside, young children are encouraged to germinate runner bean or pea seeds in jam jars. In later years, planting seeds may well be an annual event in order to raise plants in the garden or window-box. For the commercial grower, farmer or horticulturalist, seeds are the investment made in the hope of a crop in the future, and even for those of us who do not deal with seeds on that scale, the image of a farmer sifting grains of 'corn' through his fingers is readily conjured up.

Seeds are mysterious, because despite the familiarity arising from handling seeds, the way seeds 'work' is unknown to many people. We may not now think that a seed 'dies' when it is planted in the ground, but the processes occurring between the time of planting and the time the 'seed comes up' are nevertheless not widely known. Equally mysterious to many people are the processes leading to the formation of the seed on the parent plant.

For the plant physiologist, seeds are a fascinating and exciting subject of study. The process of seed development, whereby a tiny plant is formed from a fertilized egg cell is enough to occupy a life-time of research. Within the fertilized egg cell is the genetic information which directs, firstly, the essential polarity of the plant, then cellular organization and differentiation leading to the formation of a recognizable plant with defined organs, and after germination, to the form of the 'adult' plant. We have some superficial knowledge of some of the intrinsic and external factors which regulate these processes, but the details of what type of gene might control, for example, the shape of a leaf, or how an individual gene is switched on or off are unknown.

Although a seed contains a recognizable miniature plant, its development is arrested by the process of dehydration or ripening. The desiccated embryo is a truly remarkable structure: a relatively well-developed complex multicellular organism, with a moisture content of 10–15%. In this desiccated state, the seed is well adapted as a dispersal unit, and further, the widespread occurrence of seed dormancy in wild plants, coupled with specific requirements to break dormancy help to ensure that the seed germinates in a suitable habitat. The physiology of dormancy and germination is thus linked with the ecology of the parent plant.

Once germination is initiated, the growth of the embryo, which had been arrested by desiccation, re-starts. The growing regions of the embryo, the root and shoot, begin growth again where they left off during seed ripening. However, for other parts of the seed, such as specialized storage organs or tissues, germination reverses some of the processes which occur during seed development. In particular, the storage reserves, including protein, carbohydrate and/or lipids, laid down in development, are hydrolysed during germination in order to provide the growing embryo with nutrients, until it establishes itself as an independent seedling.

All these processes would be exciting and interesting had they only academic relevance. However, seed physiology is also of immense practical importance. A significant proportion of the world's population relies heavily on seeds, particularly those of cereal crops and to a lesser extent those of legumes (e.g. beans), for food. Even in those regions of the world where there is less direct reliance on seeds, seeds and seed products such as flour and oils are of considerable nutritional importance. Consequently there are ongoing efforts by plant breeders, and very recently by genetic engineers, to produce crops with higher yields of seed with improved nutritional quality or commercial value.

The five chapters of this book expand the major themes mentioned above. Before proceeding to these chapters, a digression, to explain the nomenclature used, is necessary. In order to avoid making the text too cumbersome, I have not used the systematic or Latin names for crop plants whose English names are widely known and understood, except where a Latin name is necessary for clarity. For wild plants, however, where there is a lot of local variation in English names, I have given the Latin in addition to the most widely used English name.

1 Seed Development

1.1 What is a seed?

1.1.1 Introduction

The seed is both the culmination of the activities of one plant generation and the start of a new generation. It contains within it a plant in miniature, the embryo, with the potential for growth and development into an adult plant. However, the embryo is not expressing that potential, since its development has been arrested by dehydration. In this dehydrated state, the seed is shed from the parent plant to become a unit of dispersal and eventually, under appropriate conditions, to resume growth. How then does this remarkable structure arise? In order to answer this question it is necessary to consider briefly the floral organs of the parent plant, concentrating here, as elsewhere in the book, on those of the Angiospermae (flowering plants).

1.1.2 Floral organs and pollination

The ovary of the female plant contains one or more ovules (depending on species). Each ovule contains a haploid egg cell or ovum, five other haploid cells and a central cell, containing two haploid nuclei, the polar nuclei (Fig. 1-1). During pollination, the growing pollen tube penetrates the micropyle and the two sperm nuclei enter the ovule. One fuses with the egg cell and the other with the two polar nuclei of the central cell.

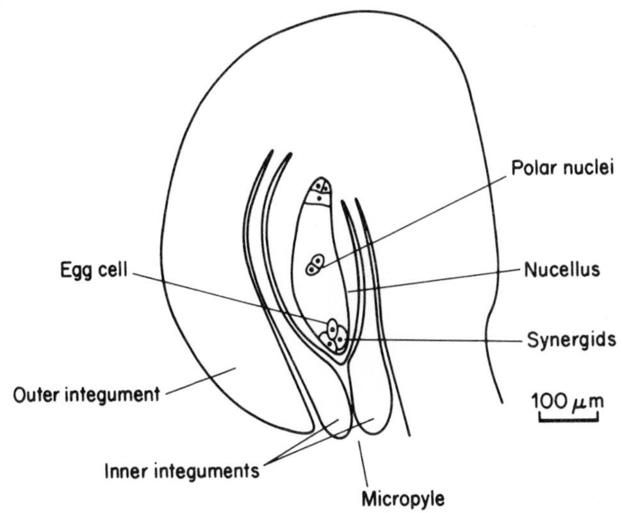

Fig. 1-1 Longitudinal section through a typical angiosperm ovule.

1.1.3 General features of seed morphology

The seed then, arises from the fertilized ovule and as the seed develops, the following seed parts are visible: *(i)* the embryo itself, which results from the fertilization of the egg cell by one sperm nucleus; *(ii)* the endosperm, which is a tissue produced as a result of the triple nuclear fusion between one sperm nucleus and the two maternal polar nuclei (in many plants this tissue becomes the major site in the seed for deposition of food reserves as, for example, in wheat, castor bean); *(iii)* the perisperm, derived from the nucellus; and *(iv)* the seed coat or testa, which is derived from the integuments of the nucellus. Having described these general features of seed morphology, three further points must be made. Firstly the extent to which the various components develop varies considerably between species (Fig. 1-2). Thus, very many mature seeds lack a perisperm. Others, such as peas and beans do not have an endosperm at maturity (the endosperm is transient and disappears during seed development); in cereals, the endosperm, although well developed, consists of dead cells in the mature seed. Secondly, there is an enormous variety in seed size. Orchid seeds, for example are tiny (in some orchids, a single seed weighs as little as one microgram with a very small embryo and almost nothing in the way of food reserves). At the other end of the scale is the double coconut which can weigh up to ten kilograms. In between these extremes are all the seeds with which we are familiar – the garden plants, crop plants and common wild

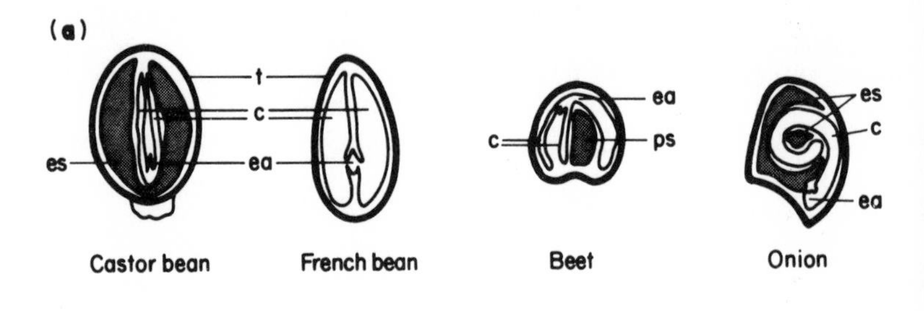

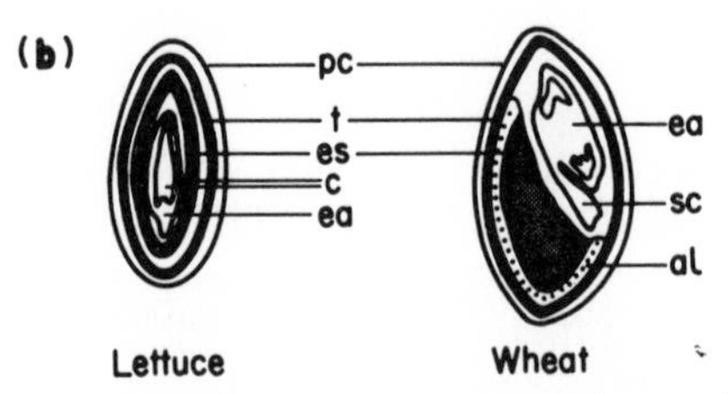

Fig. 1-2 Variation in seed morphology as illustrated by diagrammatic sections through **(a)** four seeds and **(b)** two one-seeded fruits. The individual diagrams are not to a common scale: **al,** aleurone layer; **c,** cotyledon; **ea,** embryo axis; **es,** endosperm; **pc,** pericarp; **ps,** perisperm; **sc,** scutellum (the highly modified cotyledon in cereals); **t,** testa (seed coat).

plants. Thirdly, according to the strict definition of seed structure given above, some of the structures we think of as seeds are actually one-seeded fruits: they are derived not just from the ovule, but from the entire structure which bears the ovule (the ovary or gynoecium). In these species, the outer wall is not the testa or seed coat, but the pericarp: the testa is usually reduced to a rudimentary state and may in fact be fused with the pericarp, as in cereal grains. Other examples of one-seeded fruits include lettuce (Fig. 1-2), sunflower and ash.

1.2 Embryogenesis and seed development

Seed development is the process, or more properly the wide range of processes which lead from a fertilized ovule to a mature seed, i.e. to the formation of a potentially independent plant. The range of seed morphologies has already been noted, and it will be obvious that there is also a range of developmental patterns, the only general feature of which is embryogenesis, the formation of the new plant itself within the seed.

1.2.1 Growth of the embryo

When the seed is ripe or mature, the embryo consists of an embryo axis with a shoot (plumule) and root (radicle) and one or two cotyledons, depending on the type of plant. (Note that many gymnosperms, e.g. conifers, have seeds with several cotyledons.) The cotyledons are in fact leaves and each is attached to the axis by a hypocotyl, which is equivalent to a petiole or leaf-stalk (Fig. 1-3). The embryo is thus clearly differentiated into identifiable organs and on closer inspection it also becomes apparent

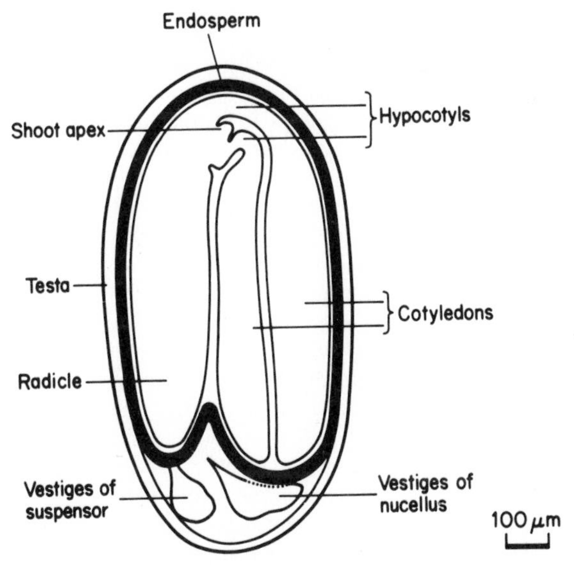

Fig. 1-3 Longitudinal section through a mature seed of Shepherd's purse, *Capsella bursa-pastoris*.

that many of the different types of cell found in the mature plant are already present in the embryo, albeit in a dehydrated, unexpanded state. So how does this little plant, the embryo, develop? This apparently simple question encapsulates what is one of the most exciting areas of biology: the control and co-ordination of cell division, gene expression and cell growth, to produce a complex multicellular organism from a single cell, the fertilized egg.

The first division of the fertilized egg results in the establishment of polarity: the top and bottom of the embryo are already distinguishable. This establishment of polarity is the first step in the development of a complex organism, and is therefore of very great significance. It is not known at all what controls polarity in higher plant embryos. In the seaweed, *Fucus*, there is clear evidence that polarity is established because of a gradient of electrical charge which arises following an asymmetric distribution of calcium ions in the fertilized egg cell. It seems probable, in view of the necessity for calcium in the regulation of cell division, that a similar mechanism may operate in flowering plants. Unfortunately it is extremely difficult to carry out suitable experiments on a single cell effectively buried in maternal tissue.

From the two-cell stage onwards there are differences in developmental patterns between species, and particularly between dicotyledons and monocotyledons (Fig. 1-4). In monocotyledons, the basal cell does not divide again, but becomes the basal cell of the suspensor. The apical cell gives rise to the rest of the suspensor and to the embryo proper. In dicotyledons the basal cell undergoes several divisions to form the suspensor (which is larger than in monocotyledons) whilst the apical cell gives rise to the embryo proper.

The suspensor used to be regarded merely as an anchorage device for the embryo, but it is now regarded as having an important role in the transfer of nutrients and plant growth regulators from the maternal tissues to the embryo. Certainly, experimentally induced damage to the suspensor upsets the growth of the embryo. In dicotyledons, the cells of the suspensor undergo DNA endo-reduplication, that is DNA replication without nuclear division. Although this is a fairly widespread phenomenon in the plant kingdom, the suspensor represents the extreme case. Some cells in the suspensor of runner bean, for example, contain 8192 C amounts of DNA (i.e. 8192 times as much DNA as a haploid cell). Assuming that these suspensor cells were in the diploid state originally, this huge amount of DNA represents twelve rounds of DNA replication in the absence of nuclear division! The function of this DNA endo-reduplication is totally unknown.

As the suspensor develops, so does the embryo, and of course embryo development continues after the formation of the suspensor is complete. As early as the 16-cell stage in dicotyledons, there is differentiation into an inner zone, consisting of cells which will form meristematic tissues, and an outer zone, consisting of cells which will form the epidermal layers. By the

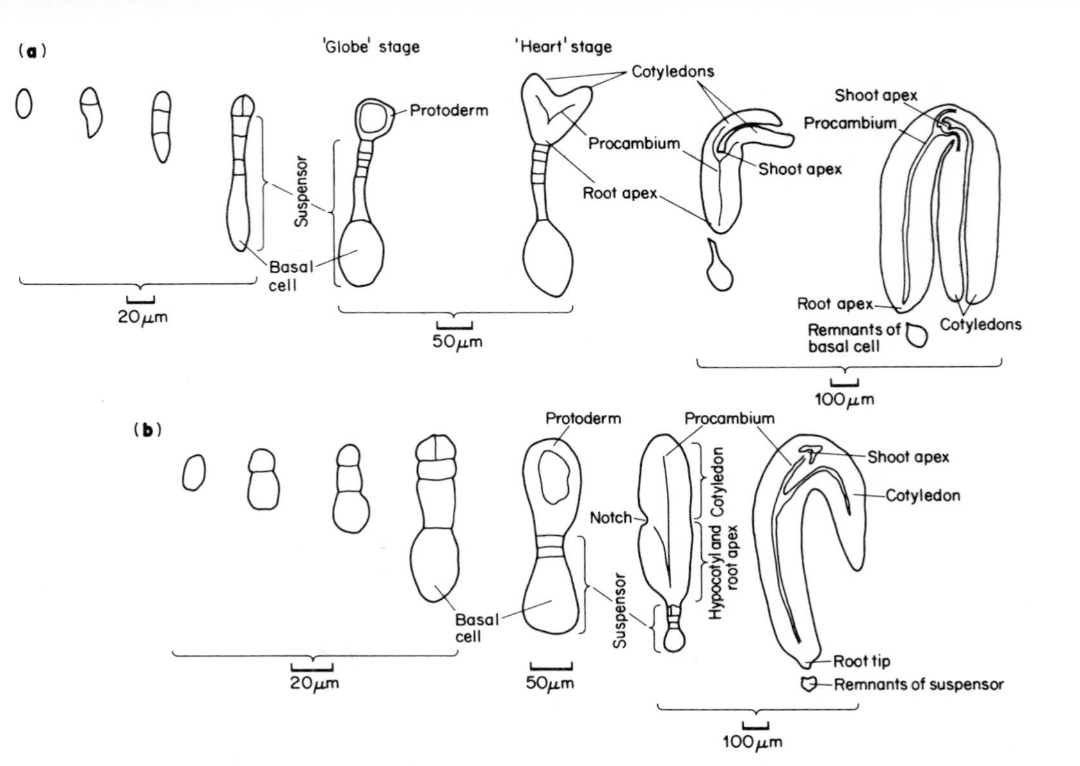

Fig. 1-4 Embryogenesis in **(a)** a 'typical' dicotyledon and **(b)** a 'typical' monocotyledon.

time the embryo contains about 50 cells, and is globular in shape (Fig 1-4), particular zones are committed to develop into particular organs, and shortly afterwards, rudimentary forms of the organs themselves, particularly the cotyledons and the root, become discernible. In dicotyledons this stage is known as the 'heart' stage, where the heart shape is caused by two distinct lobes, the rudimentary cotyledons (Fig. 1-4).

As the morphology of the embryo becomes clearer, so does the differentiation and organization of the different cell types. The dividing cells become organized into distinct meristematic regions, the epidermal and 'ground' tissues become more distinct, the cotyledonary cells, particularly in dicotyledons, take on distinct features, often associated with deposition of reserves (see Section 1.2.3) and the first stages of vascularization occur. The remainder of the changes occurring during embryogenesis (as in Fig. 1-4) simply represent an enlargement and maturation of the embryo.

The factors controlling this ordered development of the embryo are very complex. Some ideas about regulation have been obtained by excising embryos and culturing them *in vitro*, though it has not yet proved feasible to culture preglobular embryos. Unfortunately, therefore, it has not been possible to ascertain the factors leading to the commitment of different zones to particular developmental pathways. The requirement for successful culture of globular and heart-stage embryos include a high osmotic potential, a balanced nutrient medium, including nitrogen compounds, and a suitable carbon source (such as sucrose) plus the plant growth regulators auxin and kinetin. These factors are presumably normally provided by the natural environment of the embryo, which is itself surrounded by an endosperm (at least in the early stages though in many dicotyledons the endosperm is later absorbed). The endosperm has a high biosynthetic activity and also acts as one of the routes of transfer between the maternal tissues and the embryo; it would then certainly be expected to exhibit a high osmotic potential.

Later, during the maturation phase, the requirements of the embryo become less stringent. A lower osmotic potential is required, paralleling the natural situation where the endosperm is lost, as in legumes, or where the concentration of soluble compounds decreases in favour of insoluble storage compounds. The nutrient requirements become less critical, presumably as the embryo's own synthetic potential changes. There is also evidence from certain dicotyledons of a change in hormonal requirements, with, perhaps surprisingly, abscisic acid being important in the maturation phase (Section 1.2.3).

1.2.2 The endosperm

Whilst the fertilized egg cell is developing into an embryo, the triploid nucleus arising from the fusion of the two polar nuclei with one sperm nucleus (Section 1.1.2) starts to divide, giving rise to the endosperm. In many plants, for example, cereals, the early nuclear divisions occur without the laying down of cell walls, so that the endosperm is said to be coenocytic. As the endosperm enlarges it surrounds the growing embryo and, as noted above (Section 1.2.1) acts as a source of nutrients for the growing embryo. (Note that in a small proportion of species, such as sugar beet, the perisperm serves this function.) As the seed develops and the embryo within it matures, the endosperm may undergo one of two developmental pathways. Firstly, in many species, the endosperm is persistent and becomes the storage tissue for the mature seed, to be drawn on by the seedling late in germination (Chapter 3). This is seen in cereals, where in the ripe seed, the endosperm cells, although packed with reserves, are actually dead. In cereals, the cotyledon is a highly modified structure known as the scutellum, the function of which is to act as a transfer organ between the endosperm and embryo during germination (Chapter 3). Other endospermous seeds include castor bean, where the mature endosperm consists of living cells. Secondly, as typified by the legumes such as pea, and by oilseed rape, the endosperm may degenerate as the embryo matures. In such seeds, the cotyledons become the major storage organs.

1.2.3 Food reserves

Although there are some exceptions, such as the tiny dust-like seeds of orchids, most mature seeds contain a supply of nutrients consisting of carbohydrates and/or lipids as a carbon source, together with proteins and minerals (Table 1-1). These sustain the germinating seedling before it can function fully independently. As noted briefly in the previous section, the two most usual locations for these reserves are the cotyledons (part of the embryo) and the endosperm (external to the embryo). Rather more rarely, the perisperm may be the major storage tissue, as mentioned above (Section 1.2.2). The general features of the deposition of reserves are similar in the different types of storage tissue, although obviously there are differences with regard to the actual composition of the reserves and in the details of the timing of the various events in the storage process. However, the pattern is general enough for one group of plants, the legumes (which have cotyledonary reserves) to serve as a typical example (Fig. 1-5).

The differentiated but immature embryo (Fig. 1-4) arises by a rapid series of cell divisions from the fertilized egg, but then cell division activity declines and cell expansion takes over. However, although cell division ceases, DNA replication may not do so. In cotyledon cells in pea, broad bean and French bean, DNA endo-reduplication occurs. This is not quite as spectacular as that seen in the suspensor (Section 1.2.1) but nevertheless, DNA amounts of 64C are common. It is not known whether this is a general feature of seed storage tissues, although a rather limited amount of endo-reduplication has been observed in soybean cotyledons and in nuclei of the coenocytic endosperm in cereals. The function of DNA endoreduplication here, as in the suspensor, is not known. However, two theories have been suggested. Firstly, since there is a relationship between DNA amount and cell size, these very high amounts of DNA may ensure the

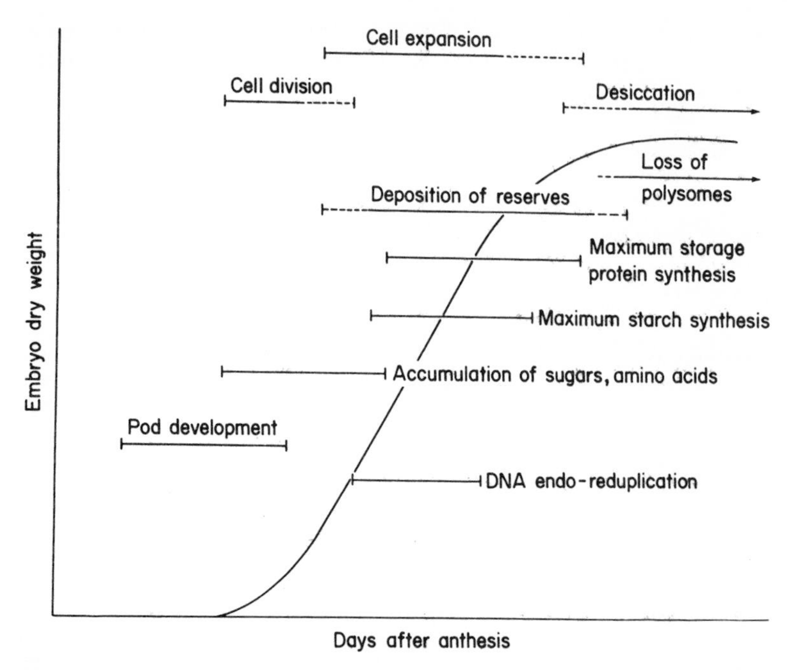

Fig. 1-5 Cotyledon development and the deposition of reserves in pea, *Pisum sativum.*

production of very large storage cells. Secondly, since a cell with a 64C amount of DNA contains 32 times as many copies of each gene as does a normal diploid cell, DNA endo-reduplication may represent a means of amplifying those genes which regulate the deposition of storage products (along with all the other genes, of course). This in turn may make for a more rapid and efficient completion of the programme of deposition of reserves.

In legumes and in other 'non-endospermic' seeds, the space left by the degenerating endosperm is filled by the expanding cotyledons; the nutrients arising from the breakdown of the endosperm are absorbed by transfer cells which differentiate in the outer epidermis of the cotyledons. As the cotyledons start to expand, the deposition of reserves is initiated. In many seeds, protein and starch are the most abundant reserves, with lipid and phytic acid (Fig. 1-6) making a small contribution. In certain seeds, such as

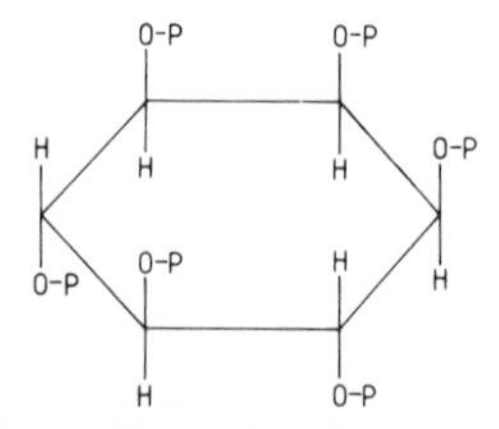

Fig. 1-6 Structure of phytic acid (*myo*-inositol hexaphosphate). P = phosphate.

soybean, peanut and rape however, lipids make up a much larger proportion of the reserves (Table 1.1). The protein and phytic acid are deposited in protein bodies which are 1–2 μm in diameter; as the protein bodies accumulate, so they replace the original vacuoles of the cell. Lipid is laid down in lipid bodies (*c.* 0.1 μm in diameter) and huge starch grains (10–20 μm in diameter) are deposited in highly modified plastids known as amyloplasts. In the fully expanded state, the storage cells are very large (*c.* 200 μm in diameter) and are so packed with storage products that the cytoplasm is confined to a thin layer at the cell periphery (Fig. 1-7).

Several of the storage products laid down in seeds do not occur anywhere else in the plant. Thus, the lipids are triglycerides and differ from the various types of diglyceride which occur, for example, in cell membranes (Fig. 1-8). Further, some of the fatty acids making up the triglycerides are found only in storage lipids, such as the ricinoleic acid in castor-bean endosperm.

Storage proteins are also highly specialized, and do not occur elsewhere in the plant (except for very small amounts in the embryo axis). In dicotyledons, the storage proteins are a small group of globulins (proteins soluble in salt-solution) which are relatively low in sulphur amino acids (see Chapter 5). Although the names given to the storage proteins differ according to species, there is in fact a large overall similarity between the storage globulins of different dicotyledons.

The deposition of large amounts of a small number of proteins means that developing cotyledons have provided a useful source of material for studying the expression of genes. The genes coding for the storage proteins occur as a gene family: instead of two copies in a diploid cell, there are several. This is obviously one method for increasing the abundance of the gene products (and the DNA endo-reduplication observed in cotyledons may be another). Further, the messenger RNA molecules copied from these genes are very abundant, perhaps to the extent of several thousand copies per cell. The genes themselves appear to be copied more frequently or rapidly than other genes, and the messenger RNA molecules appear to be

Table 1.1 Major storage compounds in seeds, expressed as % of seed dry weight.

Species	*Starch*	*Protein*	*Lipid*
Maize (*Zea mays*)	75	12	9
Wheat (*Triticum aestivum*)	75	12	2
Barley (*Hordeum vulgare*)	76	12	3
Pea (*Pisum sativum*)	56	24	6
Soybean (*Glycine max*)	26	37	17
Peanut (*Arachis hypogea*)	12	31	48
Rape (*Brassica napus*)	19	21	48

more stable than usual. Finally, the activity of these genes is absolutely confined to the embryo, and within the embryo, mainly to the cotyledons. The genes are not active anywhere else in the plant nor at any other stage of the plant's life cycle. The control of the activity is clearly at the level of the copying or transcription of the genes, since these genes are not copied into messenger RNA at all anywhere but in the developing embryo.

The cotyledons of legumes, or of rape-seed, then, are highly specialized organs developing at a particular time in the life of the plant. Further, these organs will have a totally different function during germination (see Chapter 3). What factors regulate the ordered maturation of the cotyledons during seed development? As with the younger embryos, the maternal environment has an important role to play. If French bean or rape-seed embryos are removed from the parent plant during the phase of cotyledon

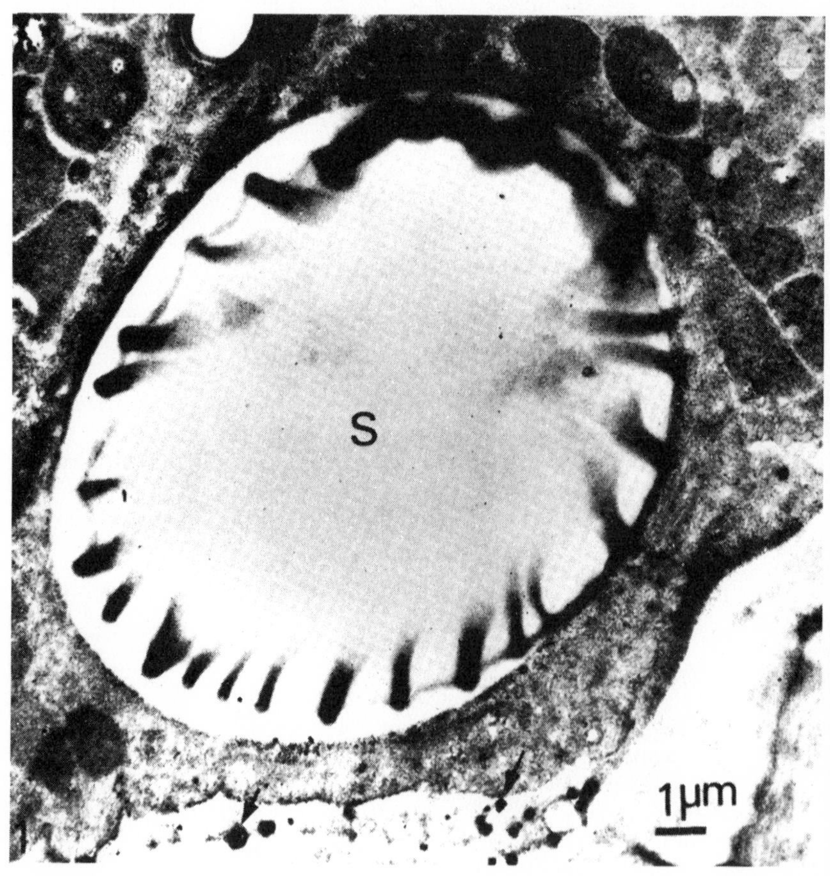

Fig. 1-7 Part of a storage parenchyma cell from a pea cotyledon. S = starch grain; P = protein body; the arrows indicate the thin layer of cytoplasm.

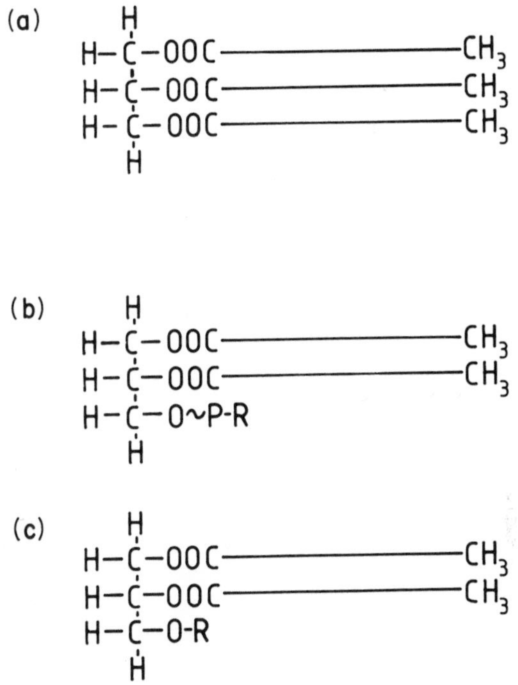

(a)

```
    H
H—C—OOC————————CH3
H—C—OOC————————CH3
H—C—OOC————————CH3
    H
```

(b)

```
    H
H—C—OOC————————CH3
H—C—OOC————————CH3
H—C—O~P-R
    H
```

(c)

```
    H
H—C—OOC————————CH3
H—C—OOC————————CH3
H—C—O-R
    H
```

Fig. 1-8 General structure of **(a)** storage lipids (tri-glycerides or triacyl-lipids) and **(b)** and **(c)** membrane lipids (diglycerides or diacyl lipids). In all three diagrams, the fatty acid chains are indicated by OOC—CH_3. In **(b)** the common substituents at R include glycerol, glycerol phosphate, ethanolamine, choline, inositol and serine. In **(c)** R is usually a sugar.

expansion and are then incubated in moist conditions, they start to germinate precociously. The deposition of reserves is halted, and eventually hydrolysis of reserves is initiated. In other words, once away from maternal environment, the developing embryo behaves as if it were mature and ripened. If the growth medium is supplemented with abscisic acid and sucrose, the excised embryo continues to behave as a developing embryo in all respects, including the regulated deposition of reserves (Table 1-2).

1.2.4 Long-lived messenger RNA

The phenomenon of precocious germination indicates that the embryo, although still developing, is also capable of germinating. In other words, there exists in the maturing embryo the potential to carry out two different phases of its life cycle. The investigation of this phenomenon in cotton seed was one of the factors which led to the discovery of long-lived messenger RNA. Cotton embryos, induced to germinate precociously, can do so even if the synthesis of messenger RNA is prevented by inhibitors (such as actinomycin). This means that the messenger RNA molecules, needed in

Table 1.2 Effect of abscisic acid (ABA) on accumulation of storage protein in rape-seed embryos during five days in culture.

ABA concentration in culture medium	*Storage protein content, μg mg⁻¹ fresh weight*
0	0.78
1×10^{-7}M	3.64
1×10^{-6}M	11.70
1×10^{-5}M	15.47

germination, are already present in the developing embryo. In some respects this is not surprising, since many of the 'working' proteins of the cell, for example enzymes, are needed both in development and germination; the messages coding for them will therefore be present at both stages. However, what is surprising is that messenger RNA molecules coding for certain enzymes involved late in germination are already present in the developing embryo. The enzymes concerned are those involved in the degradation of food reserves (see Chapter 3). Thus, at the same time as the reserves themselves are being laid down in the developing embryo, the messenger RNA molecules which will code for the degradative enzymes during germination are also being laid down. This phenomenon, first discovered in cotton, is now known to occur in many types of seed, and may indeed be universal. It is therefore possible to identify in developing embryos three populations of messenger RNA molecules: *(i)* the messenger RNA molecules coding for the working proteins of the cell, common to

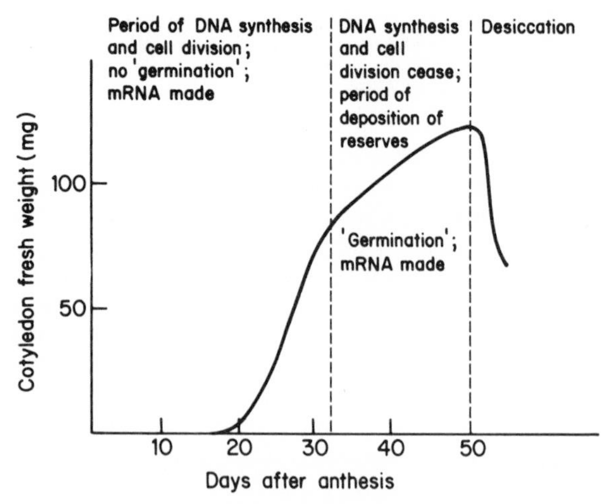

Fig. 1-9 Seed development in cotton showing time of synthesis of messenger RNA (mRNA) molecules which are used in germination.

both development and germination; *(ii)* the messenger RNA molecules specific to seed development, such as those coding for seed storage proteins; and *(iii)* the messenger RNA molecules which are stored, awaiting use in processes which are specific to germination, such as the degradation of reserves. Further, in cotton seed it has been possible to identify the time during embryo development when the population of messenger RNA molecules in *(iii)* above starts to be laid down. Very small embryos at 25 days after fertilization will germinate precociously, although very slowly. However, if messenger RNA synthesis is prevented, the specific degradative enzymes are not synthesized. This means that in these very small embryos, the messenger RNA molecules coding for the degradative enzymes have not yet been synthesized (Fig. 1-9).

The full biological significance of the ability of cells to lay down a population of messenger RNA molecules for a future stage of development is not understood (see Chapter 3). It certainly means that the control of gene expression must not *just* be assumed to be a matter of switching genes 'on' and 'off' (even if this is true of the genes coding for storage proteins).

1.3 Seed ripening and dehydration

Eventually, the embryo reaches full size, the deposition of reserves is completed and ripening or dehydration sets in. The transport of water, out of and away from what were fully turgid cells, is initially an active process. It may be aided in the later stages, either because the seed becomes detached within the enclosing fruit (and is therefore unable to take up water), or because water uptake is hindered by the structure of cells at the abscission zone. Although partly an active process, dehydration is affected by environmental conditions, particularly atmospheric moisture content. Thus, in a wet summer drying wheat grains reach a moisture content of 15% (by weight), whereas in a hot dry summer, the moisture content may be reduced to as low as 10% (however it should be remembered that even a moisture content of 15% represents extreme dehydration for living cells).

The dehydrated seed is physiologically inert or quiescent. Its metabolic activity is extremely low. No further transfer of materials from maternal tissues occurs. The seed is ready to be shed and the embryo within is ready to become an independent plant. In many respects, the seed is now ready to germinate. However, at this stage of their life, many many seeds are not merely quiescent, but dormant: they will not germinate at this time, even if exposed to appropriate conditions. This dormancy helps to ensure that the seed does not germinate should it chance to become rehydrated whilst still on the parent plant (but see Chapter 2). Earlier in this chapter it was noted that precocious germination of developing embryos was prevented at least partly by abscisic acid secreted from the maternal tissues. With mature, dehydrated embryos there is evidence from several species for an accumulation of abscisic acid within the embryo, imposing a state of dormancy, at least for a short time. The significance of dormancy, and the release from dormancy are discussed in the next chapter.

2 Dormancy

2.1 Widespread occurrence of dormancy

As indicated at the end of the previous chapter, seeds of many species may pass through a dormant state during ripening, and whilst in this state they are unable to germinate even when subjected to suitable conditions. The imposition of a state of dormancy during the late stages of ripening is obviously advantageous to the plant since it is a barrier to the germination on the parent plant of the mature or nearly mature seed. It should be noted however, that even within a given species, the depth of dormancy of the maturing seed varies from plant to plant and from year to year. This is seen in the phenomenon of cereal grains sprouting in the ear, which can occur if there is a period of particularly wet weather in the late summer.

Although the vast majority of seeds pass through this state of dormancy during ripening, most of the seeds used in agriculture or horticulture lose their dormancy either just before abscission from the parent plant or shortly afterwards (see Section 2.7.2). Thus, by the time the seeds are to be sown, they are merely quiescent ('germination-ripe' in the terminology of the seed technologist) and will germinate under appropriate conditions. This feature is clearly of great use to the grower and is a result of selection against dormancy over many generations. This is well illustrated in comparisons of crop species with closely related wild species in which there has been no selective breeding. Seeds of oat (*Avena sativa*) and runner bean (*Phaseolus coccineus*), for example, do not exhibit long-term dormancy, whereas wild oat (*Avena fatua*) and many wild species of *Phaseolus* do exhibit seed dormancy. However, selection against long-term dormancy has not been entirely successful with all vegetable and ornamental flower seeds. Amongst vegetable crops, beet is particularly well-known for its dormancy. Such dormancy in a commercial crop is a nuisance, since it makes the prediction of germination performance very difficult.

Whilst long-term dormancy is rare amongst the seeds which are used commercially, it is very widespread amongst seeds of wild plants. For example, of 403 species of British wild plants growing mainly in the vicinity of Sheffield, in South Yorkshire, 39% exhibited a high degree of seed dormancy (more than 90% of the seeds were dormant) whilst in a further 29% of species, between 50% and 90% of the seeds were dormant. Neither should it be thought that dormancy is a feature of only temperate and cool-temperate floras. Dormancy, in fact, occurs in the floras of all the climatic zones, ranging from tropical to arctic-alpine. Consideration of only crop or garden plants then would give a very unbalanced view of the occurrence and importance of seed dormancy.

2.2 Relative dormancy

As has been hinted at already, dormancy is not necessarily an 'all-or-nothing' phenomenon. For example, some seeds may exhibit dormancy at one particular temperature whilst not exhibiting dormancy at another. Wheat grains are usually dormant for a short while after harvest, and indeed will not at that stage germinate at 20°C. If however the temperature is lowered to 15°C, nearly all the grains germinate. After the loss of dormancy, germination occurs at both temperatures (Fig. 2-1). This type of seed behaviour where dormancy is exhibited under some conditions, but not under others, is called relative dormancy.

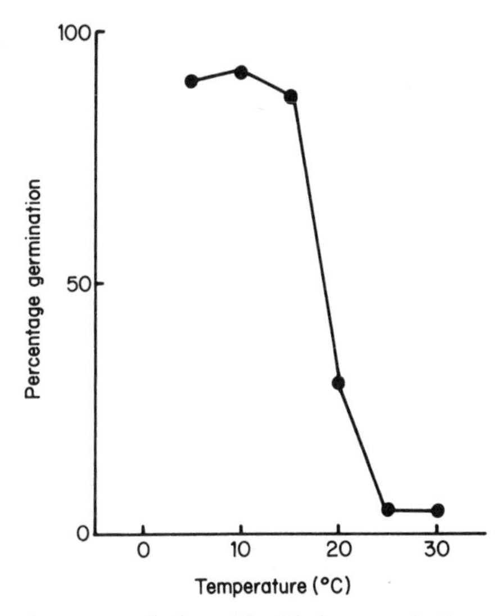

Fig. 2-1 Relative dormancy of wheat. Freshly harvested wheat grains set to germinate at different temperatures exhibited dormancy only at higher temperatures.

2.3 Secondary dormancy

There are several plant species known in which the seeds are not initially dormant. They will germinate under suitable conditions, but become dormant if exposed to conditions which are unsuitable for germination. Seeds of celery, for example, become dormant if exposed to temperatures too high for germination, whilst seeds of ivy-leaved speedwell (*Veronica hederifolia*) become dormant if exposed to temperatures too low for germination.

2.4 Causes of dormancy

Seed dormancy is imposed by a number of different mechanisms, according to species. These mechanisms may be grouped into two broad

Table 2.1 Coat-imposed and embryo dormancy in the genus *Acer*. Freshly harvested seeds of *Acer pseudoplatanus* (sycamore) and *Acer platanoides* (Norway maple) (both of which require cold treatment to break dormancy), were set to germinate either intact, or with the testa removed at 17°C. The experiment was repeated with seeds stored at 2°C for 110 days.

Species	*Seed treatment*	*Germination(%)*	
		Intact seeds	*Testa removed*
A. pseudoplatanus	Fresh	0	88
	110 days at 2°C	90	93
A. platanoides	Fresh	1	1
	110 days at 2°C	75	85

categories, embryo dormancy and coat-imposed dormancy. In embryo dormancy, some factor(s) in the embryo itself prevents germination. This may be shown by removing the seed coat and incubating the isolated embryo under conditions favourable for germination. If the dormancy is a true embryo dormancy, no embryo growth will occur under these conditions. In coat-imposed dormancy, the seed coat prevents the embryo from germinating and so removal of the seed coat, followed by incubation, will result in embryo growth. The difference between these two basic categories of dormancy is illustrated in Table 2-1. It should be noted that in some species, dormancy of both categories is exhibited, as in hawthorn (*Crataegus*). It should also be noted that in some instances, the difference between the two categories is not clear-cut (see Section 2.9). Further, even within one broad category, it should not be assumed that only one factor operates. In charlock (*Sinapis arvensis*) for example, the seed coat, with its layer of mucilage, prevents easy access of oxygen to the embryo, maintaining a very low oxygen tension at the embryo surface. Thus, exposure of seeds to high oxygen tensions is at least partially effective in breaking dormancy. However, measurement of the ability of isolated embryos to germinate indicates that germination is possible, although very slow, at the oxygen tensions which normally occur inside the intact seed. Here then, it is likely that dormancy is a result of a low growth potential caused by low oxygen tension and a purely mechanical restraint, also imposed by the seed coat.

2.5 Breakage of dormancy

Just as causes of dormancy are diverse, so are the factors involved in breakage of dormancy. In discussing these factors, care must be taken to distinguish between the factors themselves and their underlying effects. Thus, a number of different environmental factors, such as high or low temperature or light, may lead to a breakage of seed dormancy. In some species, breakage of dormancy may follow exposure to either of two different environmental factors (such as low temperature or light). However, it must not necessarily be assumed that in such seeds there are two different internal mechanisms for breaking dormancy: it is possible that the two different environmental triggers both act on the same internal mechanism (see Section 2.8).

2.6 Environmental factors involved in breakage of dormancy

2.6.1 Light

The seeds of many wild plants are small and easily buried, for example, thyme (*Thymus praecox*), nettle (*Urtica dioica*) and groundsel (*Senecio vulgaris*). In many such seeds there is a light requirement for breakage of dormancy and the ability to respond to light depends on imbibition (i.e. uptake of water by the seed). The response also depends on the release from the 'ripening dormancy' (Chapter 1 and Section 2.1 of this chapter), so that newly shed seeds which imbibe water may not necessarily germinate even in the light.

2.6.2 Low temperature

Seeds of many species of plants which grow in temperate and cooler climates require a period of chilling, again whilst in the imbibed state, for breakage of dormancy. In general, there is a correlation between climate and the length of the chilling period required (see Chapter 4). Figure 2-2

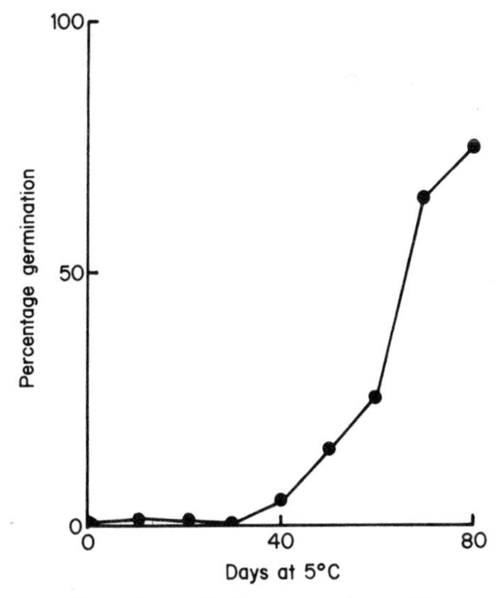

Fig. 2-2 Breakage of apple seed dormancy by cold treatment. Percentage germination at 25°C after storage for varying lengths of time at 5°C.

shows the breakage of dormancy in apple seeds by storage at 5°C. A particularly interesting example of a chill requirement is seen in the beautiful North American plant, *Trillium*. In this plant, the dormancy of the radicle (seed root) is broken by one long exposure to cold, so that germination in the sense of radicle emergence takes place after one winter. The seedling becomes established in the ground, drawing on the reserves in the seed, and on soil-borne water and nutrients, but the shoot does not emerge. Breakage of dormancy in the plumule requires a further period of chilling, and so shoot growth is not initiated until two winters have passed.

For some species, the period of chilling must be preceded by an exposure to warmer conditions. Typically such plants are spring flowering shrubs and trees, such as hawthorn. The seeds are shed in summer, exposed to high summer temperatures and then to low winter temperatures before germinating the following spring.

2.6.3 High temperature

A requirement for high temperature to break dormancy is rather rarer than a chilling requirement. Normally high temperature requirements are exhibited by early flowering 'winter' annuals: those plants which germinate in the autumn and flower the following spring, before the heat of high summer. This strategy is relatively rare in Western Europe, but is quite common in areas such as South-eastern USA, where the winters are mild and the summers are hot. The best example in Western Europe of a heat requirement is seen in bluebell (*Hyacinthoides non-scripta*). Although not a winter annual, its seeds are shed in early summer and do not germinate until they have been exposed to the heat of high summer. Germination occurs in the autumn, and the seedling is then poised to grow in the spring before the woodland canopy closes over.

2.6.4 Fire

Many shrubs and trees of sub-tropical and semi-arid regions have extremely hard seeds in which the coat is very impervious to water. Dormancy in such seeds is clearly coat-imposed, and may be broken by exposure to extreme heat such as would occur in a scrub fire. There have also been suggestions that seeds of certain European plants, such as heather (*Calluna vulgaris*), normally have their dormancy broken by fire, again because the heat cracks the seed coat. However, in heather, dormancy may also be broken by long-term changes in the seed coat as the seed lies in damp soil.

2.6.5 Alternating temperatures

A larger number of plant species which grow in temperate and cool-temperate regions have seeds which require alternating temperatures for breakage of dormancy. Bull-rush (*Typha*) and many species of dock (*Rumex*) are examples. The response is often very difficult to quantify, since so many variables are involved, including the particular temperatures, the magnitude of the fluctuation, the length of exposure to each temperature and the number of cycles. Generally, it takes only a few cycles of the optimum temperature fluctuation to break dormancy. The adaptive significance of this is discussed further in Chapter 4.

2.6.6 Water

Frequently seeds lie dormant in damp soil in an imbibed state. This allows leaching from the seed coat, or from the embryo itself, of water-soluble inhibitors. The leaching effect is very much increased during rainfall when there is an actual throughput of water through the soil. At its most spectacular this is seen in desert ephemerals, the seeds of which normally remain dry as they lie dormant. A heavy rainstorm will lead to water uptake by the seeds, and leach the inhibitors out of the seeds.

2.6.7 General environmental effects

The previous sections have identified a number of specific environmental effects which lead to the breakage of seed dormancy. However, for many dormant seeds, breakage of dormancy is just a matter of exposure. In species where dormancy is imposed simply by a very hard seed coat, lying in damp soil, particularly in fluctuating temperatures, may eventually lead to a softening of the seed coat by chemical hydrolysis and perhaps also by microbial action. Such effects are likely to be relatively unpredictable even within a given species (as anyone who has tried to grow sweet peas will know), and germination of the seeds will then be spread over a period of time.

2.7 Embryo physiology and breakage of dormancy

2.7.1 Maturation of the embryo

A special case of embryo dormancy is seen in seeds which exhibit immaturity of the embryo. In such seeds, the embryo needs to develop further, after shedding, to a state where it is ready to germinate. This can only happen when the seed is imbibed, and for the majority of seeds exhibiting embryo immaturity (such as spindle-tree *Euonymus*), the best temperatures for embryo maturation are those which support normal growth. An exception to this is hogweed (*Heracleum sphondylium*), in which embryo maturation occurs much faster at low temperatures (below 5°C) than at normal growth temperatures.

Ironically, in some species, when this post-abscission embryo maturation has finished, the matured seed is dormant in the more conventional sense and needs to be subjected to one of the environmental factors mentioned in Section 2.6. Seeds of black ash (*Fraxinus nigra*), for example, require a chilling treatment after the embryo has matured.

2.7.2 After-ripening

Some seeds, which are dormant when shed from the parent plant, gradually lose their dormancy if they are kept in dry conditions. Seeds of wild oat (*Avena fatua*), for example, lose their chilling requirement if kept in dry condition for 30 months. In ash (*Fraxinus excelsior*), the chilling requirement is partly abolished by after-ripening. The relative dormancy (Section 2.2) exhibited by light-requiring cultivars of lettuce (such as 'Grand Rapids') is also abolished by 18 months dry storage. Maintenance of seeds at low moisture content is a relatively rare occurrence in nature, although it obviously does happen in regions where rainfall is scarce. However, seed storage under dry conditions is normal horticultural and agricultural practice. This period of dry storage may well serve to remove or reduce the residual 'ripening dormancy', as exhibited, for example, in cereals (Section 2.1).

2.8 Mechanisms involved in the release from embryo dormancy

It has already been noted that it is important to distinguish between the factors which bring about breakage of dormancy and the internal mechanisms triggered by those factors. This will become clearer in the next three sections, which deal with the effects of two different environmental factors, light and chilling, and several different features of embryo physiology.

2.8.1 Light and phytochrome

Many wild plant species have seeds which require exposure to light for breakage of dormancy. However, the best researched example is the 'Grand

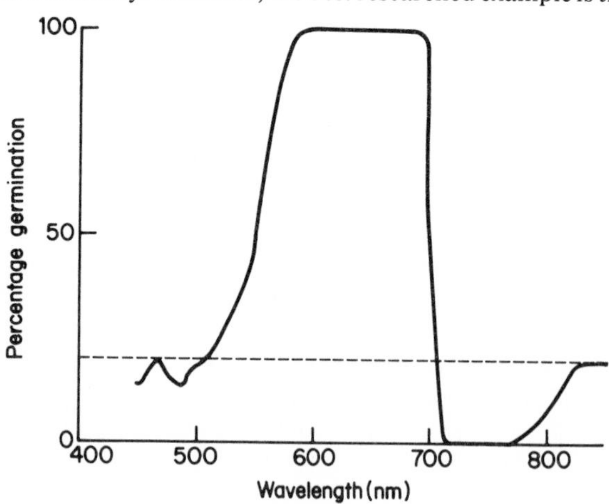

Fig. 2-3 Effects of light of different wavelengths on germination of imbibed 'Grand Rapids' lettuce seeds. The dashed line indicates the % germination of seeds kept in darkness.

Rapids' cultivar of lettuce, which has a prominent place in the history of plant physiology. When imbibed 'Grand Rapids' lettuce seeds are exposed to light of different wavelengths, it is found that light in the red region of the spectrum is the most effective in breaking dormancy (Fig. 2-3). To be more specific, light of wavelength 660 nm has been shown to be particularly effective. Figure 2-3 also shows two other important features. Firstly, a small percentage of the seeds germinate in the dark. This is a common feature in light-requiring seeds, namely that a certain proportion are not dormant (also see Chapter 4). Secondly, in the context of the experiment shown in Fig. 2-3, the presence of some seeds which do not require light helps to show that some wavelengths of light, rather than breaking dormancy, actually promote it and prevent the germination of hitherto non-dormant seeds. These wavelengths are in the far-red region of the spectrum.

Data such as these led to the discovery of phytochrome, a pigment which exists in two main forms, Pr and Pfr. The form known as Pr is *receptive* to red light. On illumination with red light, Pr is converted to Pfr, a form which is receptive to far-red light, as summarized in the following equation.

$$\text{Pr} \underset{\text{far-red light}}{\overset{\text{red light}}{\rightleftharpoons}} \text{Pfr}$$

In the dormant lettuce seed, the phytochrome is mostly in the Pr form. When illuminated with red light, the phytochrome is converted to the Pfr form which promotes the germination of the seed (by a mechanism not yet understood). However, when phytochrome is in the Pfr form, it is receptive to far-red light which converts it to Pr, and in this form it does not promote the breakage of dormancy. This can be confirmed by illuminating imbibed seeds first with light of one wavelength and then of the other (Table 2-2). Note that white light promotes germination because it contains more red than far-red light. Those seeds which are not actually dormant are thought

Table 2.2 Effects of illumination on germination of imbibed 'Grand Rapids' lettuce seed.

Illumination conditions	*Seeds germinated* (%)
Dark	20
White light	92
Red light	98
Far-red light	1
Red, then far-red	2
Red, far-red, red	98
Red, far-red, red, far-red	1

to have a lower Pr/Pfr ratio than the dormant ones. Exposure of these non-dormant seeds to far-red light will obviously raise the Pr/Pfr ratio and render the seeds dormant.

2.8.2 Plant growth substances

Plant hormones, or plant growth substances, are involved in the regulation of many aspects of plant growth and development. It was natural therefore that seed physiologists should investigate the possibility that plant growth substances are involved in seed dormancy and/or its breakage. Table 2-3 shows the effect of exposing 'Grand Rapids' lettuce seeds to the plant growth substance, gibberellic acid (GA), in the dark. It is obvious that the GA is as effective as light in breaking dormancy. On the other hand, if abscisic acid (ABA), is applied at the same time, then GA does not break the seeds' dormancy. A very similar situation exists in seeds which require chilling to break dormancy. For example, dormancy of hazel (*Corylus avellana*) seeds is broken by the application of GA; this breakage of dormancy is prevented by ABA. The inhibitory effects of ABA recall the role of this growth substance in preventing precocious germination and possibly also in the short-term ripening dormancy (Chapter 1).

Table 2.3 Effects of plant growth substances on germination of imbibed 'Grand Rapids' lettuce seeds in the dark.

Growth substance added to germination medium	*Seeds germinated* (%)
None	20
Gibberellic acid, 10^{-5}M	97
Abscisic acid, 10^{-5}M	2
GA, 10^{-5}M + ABA, 10^{-5}M	21

The plant growth substances thus provide some sort of linkage, in the breakage of dormancy, between two different factors, light and chilling, particularly in those species where either of these factors is effective. A theory which brings all these data together is that the breakage of dormancy is brought about by a change in the balance between inhibitory plant growth substances such as ABA and promotory growth substances such as GA. This could happen either by a decrease in the amount of ABA or an increase in the amount of GA, or both. It is certainly true that chilling leads to a decrease in the levels of ABA in seeds of hazel and of Norway maple (*Acer platanoides*) (Fig. 2-4).

However, it is not known how light, acting via phytochrome, influences the levels of growth substances, although there is some evidence that phytochrome affects the synthesis and transport of gibberellic acid.

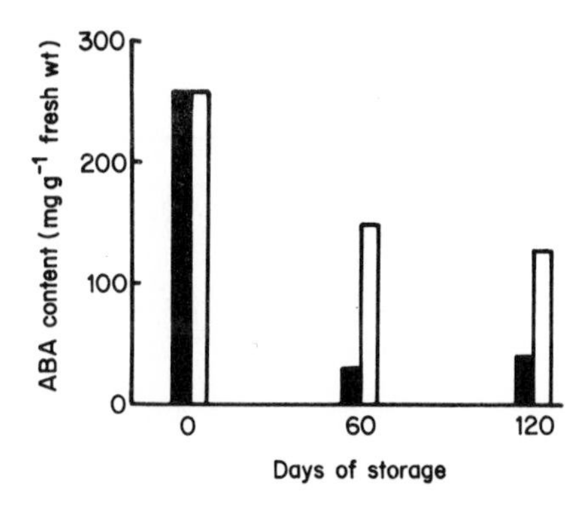

Fig. 2-4 Abscisic acid (ABA) content of embryos in seeds of Norway maple (*Acer platanoides*) stored at 5°C (solid bars) or 17°C (open bars).

Likewise, as with a number of effects of growth substances, the mechanism whereby they actually promote or break dormancy is not known. Changes in membrane permeability (and hence in ion movements) have been implicated in the action of growth regulators, and this suggestion provides a further possible common feature in the action of phytochrome and of chilling.

2.8.3 Changes in gene expression

The actively germinating seed is clearly very different from the dormant seed in growth behaviour. Does this change of state reflect changes in the expression of different genes? In several types of seed where extensive chilling is needed to break dormancy, large amounts of RNA accumulate during the chilling period, whereas RNA does not accumulate in control seeds held at higher temperatures (Fig. 2-5). These data actually hide quite a complex situation, because RNA is actually *synthesized* faster in unchilled seeds than in chilled seeds, but it *accumulates* in chilled seeds. The obvious implication here is that in unchilled seeds, the RNA is broken down as fast as it is made. This emphasizes the importance of the turnover of macromolecules in the control of metabolism, growth and development.

What, then, does the accumulation of RNA in chilled seeds represent? In hazel and Norway maple seeds it is known to be mostly ribosomal RNA; there is certainly no preferential accumulation of messenger RNA. Indeed, in the early stages of chilling, there is actually a decrease in the amount of messenger RNA, presumably caused by the degradation of the messenger RNA molecules used in seed development, but no longer needed (Chapter 1). Further, in Norway maple, there appears to be no new type of messenger RNA synthesized during chilling. The general accumulation of RNA in this species may be seen as an expression of the embryo's increasing growth potential. In hazel seed (induced to break dormancy by exposure to GA, rather than by chilling) there is some evidence for the presence of new types

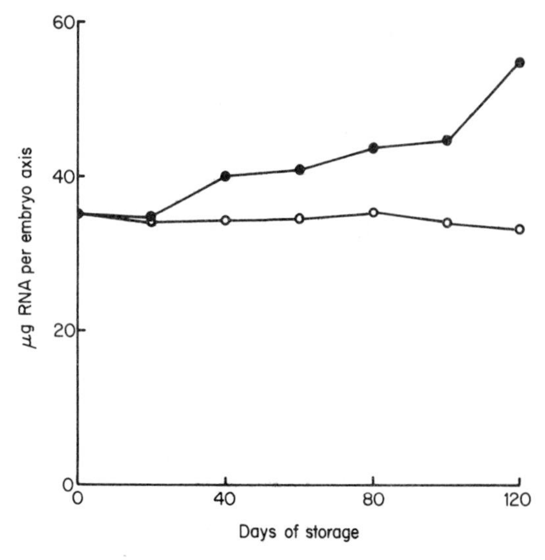

Fig. 2-5 Changes in RNA content of the embryo axes of Norway maple (*Acer platanoides*) seeds stored at 4°C (•) or 17°C (o).

of messenger RNA, and it is possible that there are differences between species in this respect.

One final point must be made here: although chilling brings about an accumulation of RNA, particularly of ribosomal RNA, it is not known whether this is really necessary for breakage of dormancy. In light-requiring seeds, for example, the stimulus may need to be presented only for a few minutes to promote germination. There is certainly no time for long-term accumulation of RNA before germination starts. Indeed, any accumulation of RNA which occurs in such seeds is a feature of germination, rather than breakage of dormancy. The behaviour of individual species, where dormancy can be broken either by a relatively long exposure to chilling or to a very short exposure to light, is particularly hard to understand in this respect. Certainly it appears here that an accumulation of large amounts of RNA is not a prerequisite for breakage of dormancy.

2.9 Breakage of dormancy: concluding remarks

The environmental triggers which bring about the breakage of embryo dormancy may also bring about the breakage of certain types of coat-imposed dormancy. Coat-imposed dormancy in sycamore (*Acer pseudoplatanus*) for example, is broken by chilling, as is embryo dormancy in Norway maple (*Acer platanoides*). So, embryo dormancy may not be qualitatively different from the type of coat-imposed dormancy where the growth of the embryo is merely physically restrained by the coat. In both, breakage of dormancy may simply be a reflection of increased growth potential in the embryo.

Other than this, it is very difficult to make any unifying statements about breakage of dormancy. The variety of environmental triggers which break dormancy in different, or even in the same seeds shows that there is not a universal perception mechanism. It is possible to unify the effects of different environmental factors by assuming that ultimately, (after perception) they act on similar internal mechanisms. The concentrations and distribution of plant growth regulators may be a common feature, and changes in membrane permeability, and hence in movement of solutes and ions, may be another. There is clearly scope here for much more research.

3 Germination

3.1 Definitions

The definition of the term 'germination' is more difficult than would appear at first sight. Many laboratory workers take the protrusion of the radicle – the seed root – through the testa (seed coat) as the culmination of germination. Gardeners and farmers on the other hand speak of seeds germinating when the shoot makes its appearance above the ground. Neither of these definitions is entirely satisfactory, since both focus attention specifically on the growing parts of the seed, and draw attention away from other, equally important parts, such as food reserves. A more general view, and one which is adopted here, is that the term 'germination' covers all the processes which are involved in the transformation of a plant embryo into an independent, established seedling. Such a definition thus encompasses processes such as the hydrolysis of food reserves, which may continue for several days (or even weeks, as in coconut) after root or shoot emergence.

3.2 Styles of germination

As already discussed in Chapter 1, there is an enormous variety in seed morphology, seed size and in the size of the food reserves. A similar variety exists in the styles of germination exhibited by different seeds. The tiny seeds of orchids, for example, cannot germinate on their own, but need to enter into a symbiotic relationship with a fungus. At the other end of the scale, the coconut seed has a huge food reserve which completely dwarfs the embryo and sustains it for a very long time after it has emerged. With such a range, it is not possible to cover all modes of germination. Instead, attention will be paid to a small number of familiar types of seed: peas and beans; marrow (squash); onion and barley. Table 3-1 summarizes the germination behaviour of these different types. Specific features of some of these are also shown in Fig. 3-1.

The six examples discussed in Table 3-1 and Fig. 3-1 show that there is enormous variety in the manner of germination, and in particular, in the role and behaviour of different parts of the seed, such as endosperm and cotyledons. It is clear therefore that care must be taken in making generalizations about germination. This will become particularly apparent in Section 3.4. However, there are a number of processes which are essential in any of the styles of germination. These are the basic physiological (as opposed to morphological) processes, and are described in the next section.

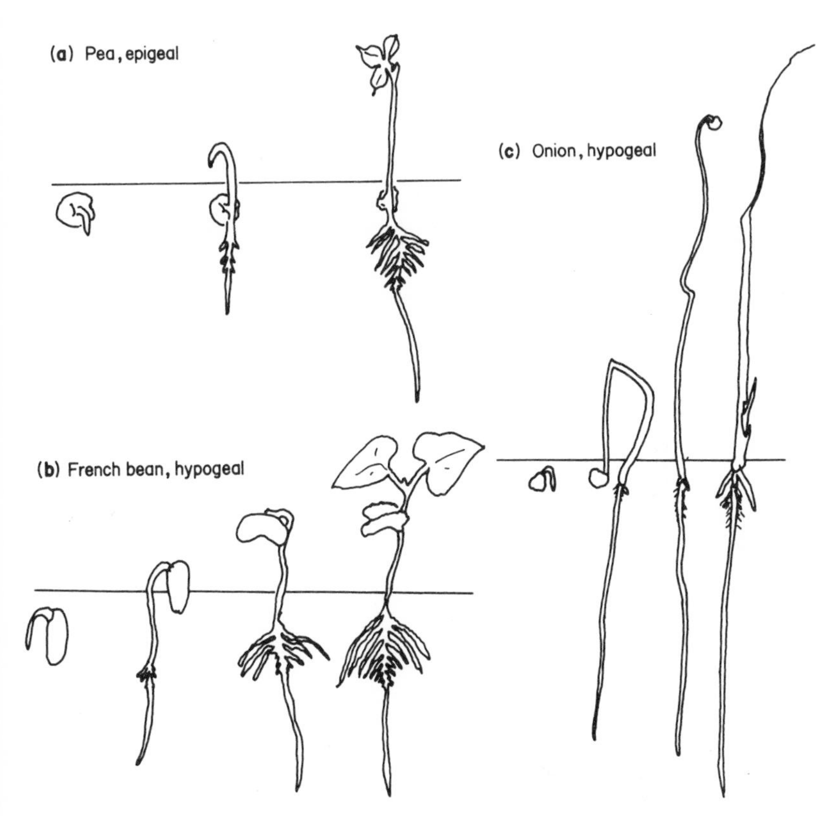

Fig. 3-1 Different styles of germination. See also Table 3.1. Note the behaviour of the cotyledons: epigeal germination, cotyledons remain below ground; hypogeal germination, cotyledons appear above ground.

3.3 General physiological features of germination

3.3.1 Rehydration, re-assembly and re-activation

The types of seed whose germination behaviour is most widely known are mainly those seeds which are used in agriculture and horticulture. As previously noted (Chapter 2), only a few commercially used species exhibit any kind of dormancy in the ripe seed. Non-dormant seeds are merely quiescent when they are dehydrated, and germination starts as the seed becomes rehydrated by the process of imbibition (water uptake). It should be noted that this is not necessarily true of seeds which exhibit dormancy, since such seeds often undergo breakage of dormancy in the rehydrated state, and it is thus difficult to determine with any precision the point at which the seeds start to germinate. In this section, then, attention is paid specifically to those seeds in which imbibition initiates germination.

Table 3.1 Styles of germination.

Species	*Endosperm*	*Cotyledons*	*General features of germination*
Dicotyledons			
Pea (*Pisum sativum*)	Non-existent in mature seed	Form main storage organs	*Hypogeal. Cotyledons shrivel as reserves are hydrolysed
French bean (*Phaseolus vulgaris*)	Non-existent in mature seed	Form main storage organs	*Epigeal. Cotyledons shrivel as reserves are hydrolysed
Marrow (squash) (*Cucurbita pepo*)	Consists of one thin layer of cells in the mature seed	Form main storage organs	Epigeal. As hypocotyl starts to elongate, cotyledons are initially still enclosed in testa. Reserves in the cotyledons are hydrolysed after shedding of testa. Cotyledons then expand and become photosynthetic; remain functional for several weeks
Castor bean (*Ricinus communis*)	Major storage tissue	Embedded in the endosperm	Epigeal. As hypocotyl elongates, cotyledons remain partially embedded in endosperm and partially enclosed in testa. Cotyledons absorb hydrolysis products from endosperm. Testa is shed and cotyledons expand to become photosynthetic. Remain functional for several weeks

continued

Table 3.1 *continued*

Species	*Endosperm*	*Cotyledons*	*General features of germination*
Monocotyledons			
Onion (*Allium cepa*)	Major storage tissue	Folded, with its curled tip embedded in the endosperm	Initial elongation is at base of cotyledon, pulling entire embryo, except for tip of cotyledon, clear of testa. Germination is epigeal. Tip of cotyledon remains buried in endosperm (where it acts as transfer organ), and enclosed by testa. Cotyledon becomes photosynthetic for a few days. After hydrolysis of endosperm is complete, testa is shed and cotyledon withers
Barley (*Hordeum vulgare*)	Consists of *dead* cells: is the major storage tissue	Highly modified (scutellum). Lies between the endosperm and the embryo axis	Hypogeal. Scutellum acts as transfer organ during hydrolysis of endosperm reserves

*See Fig. 3.1

In the quiescent seed, normal metabolic processes are held in check because of the lack of water. Imbibition initiates an almost immediate resumption of these metabolic processes. The ability to achieve an immediate resumption of metabolic activity after extreme desiccation is a very rare feature in higher plants, and apart from a small number of highly specialized plants, adapted to dry habitats, is confined to seeds. This immediate renewal of metabolic activity in seeds is sometimes not apparent to those seed physiologists working with whole seeds or whole embryos. Early in imbibition, cells near the surface are completely rehydrated, whilst cells further away from the surface may still be dry. Measurement of the metabolic activity of whole seeds or whole embryos will, therefore, indicate a gradual renewal of metabolic activity during the hours taken to complete the process of water uptake. However, the techniques of autoradiography, microscopy, and electron microscopy, which allow investigation of individual cells within a seed, show that protein and RNA synthesis, for example, are initiated within a cell almost as soon as that cell has been rehydrated.

The ability to re-initiate metabolism, more or less instantly, indicates that the components necessary to sustain metabolism (e.g. enzymes, membrane system) must survive the dehydration-rehydration process. There is, in fact, clear evidence that a mature, dry, quiescent seed does indeed contain the whole range of metabolic and synthetic apparatus required for an immediate renewal of metabolic activity. Respiration, for example, may be re-initiated before the synthesis of new respiratory enzymes; protein synthesis can get under way without the need to await the arrival from the nucleus of newly-synthesized RNA molecules. Protein synthesis is a particularly interesting example of metabolic readiness, and is dealt with in more detail in a later section (3.3.4).

Although there is an immediate renewal of metabolic activity on rehydration, it appears that metabolism in the early stages of germination is not as efficient as normal. Respiration provides a good example of this. Mitochondria extracted from pea seeds early in imbibition exhibit a lower respiratory control ratio – less adenosine triphosphate (ATP) is made per unit of oxygen taken up – than usual. This suggests that the metabolic apparatus, although present, is impaired in some way. This example leads on to the consideration of the possibility that cellular components may be damaged by the dehydration-rehydration cycle.

The idea that damage may occur in seeds comes not only from this observation of poor respiratory control in mitochondria, but also from observations that membranes are often 'leaky'. Both these observations may be explained by an incomplete reformation of the lipid bilayer of membranes, or of incomplete or incorrect re-alignment of proteins in the lipid bilayer on rehydration. There is also evidence that the genetic material, DNA may be damaged, and that RNA molecules, particularly ribosomal RNA molecules, may be broken. This is shown in Fig. 3-2. It appears then, that although the dehydration-rehydration cycle is a normal

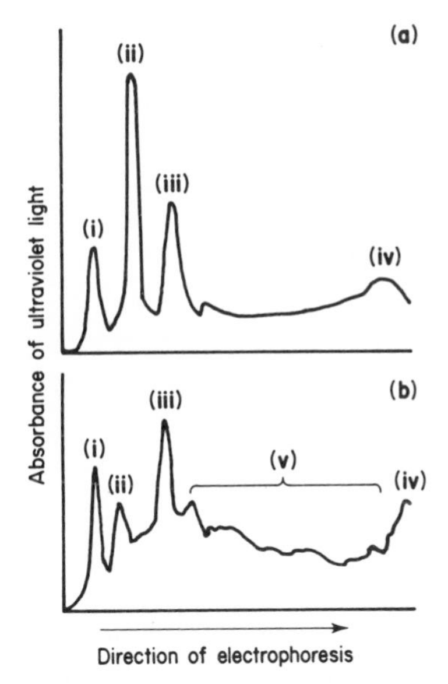

Fig. 3-2 Nucleic acids extracted from carrot seeds and separated by electrophoresis in polyacrylamide gels. The nucleic acids are detected by exposing the gel to ultraviolet light, and are seen to separate according to molecular weight: the smallest molecules travel furthest. **(a)** Electrophoresis under standard conditions; **(b)** electrophoresis under denaturing conditions. **(i)** DNA, **(ii)** 26*S* ribosomal RNA, **(iii)** 18*S* ribosomal RNA, **(iv)** transfer RNA and low-molecular weight ribosomal RNA. In **(a)** RNA molecules are complete because of internal hydrogen bonding, in **(b)** where the bonding is broken the RNA molecules appear as fragments **(v)**, with a concomitant reduction in the amount of the native molecules, particularly the 26*S* RNA.

part of higher plant life, the plant has to sustain some 'physiological cost' in going through this cycle.

So, at least some of the metabolic activity which occurs during early germination is directed at repairing or replacing damaged components. The efficiency of mitochondrial ATP synthesis is restored as components of the inner mitochondrial membrane are synthesized, and as electron carrier proteins are made and inserted into the membrane. Membrane systems in general become less leaky as phospholipid membrane components are synthesized. Gaps and breaks in DNA molecules are filled and sealed by the action of the enzymes DNA polymerase and DNA ligase, and broken RNA molecules are degraded and then replaced by newly synthesized undamaged molecules (Fig. 3-3).

3.3.2 *Renewed growth*

Although some of the metabolic activity during the early stages of germination is related to repair and replacement processes, ultimately, the

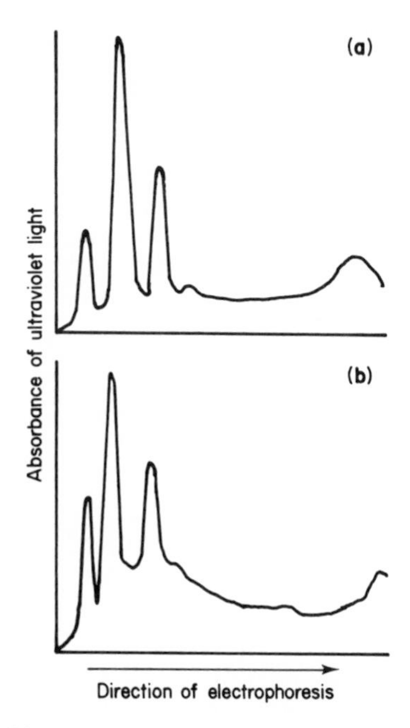

Fig. 3-3 Nucleic acids extracted from carrot seedlings and fractionated as described in Fig. 3-2.

renewal of metabolic activity is directed towards growth. Different parts of the embryo start growing at different times after the onset of imbibition. In many species, radicle growth precedes epicotyl growth by many hours, and much of the experimental work on renewed growth has been carried out on the radicle. The larger seeds, such as peas and beans are easier to work with than small seeds, and so naturally, more work has been done on these large seeds. In this section then, most of the examples concern growth of the radicle in large-seeded species, particularly the pea.

Renewal of growth is often first detected by an increase in the fresh weight of the growing part of the embryo (i.e. the embryo axis). In pea, for example, the fresh weight of the embryo axis increases markedly during imbibition and then remains constant for some 18 to 20 hours (Fig. 3-4). However, at about 22 hours (in peas grown at 22–25°C) after the onset of imbibition, the fresh weight again starts to increase dramatically. This is accompanied by cell extension in the radicle and by differentiation of the vascular tissue. Sometime between 30 and 40 hours after the onset of imbibition, the elongating radicle splits the testa and emerges. It is important to note that in the pea, and indeed, in all species investigated, this initial growth of the radicle is brought about by elongation of pre-existing cells. Cell division does not usually recommence until after emergence.

Ultrastructurally, the increase in fresh weight is accompanied by

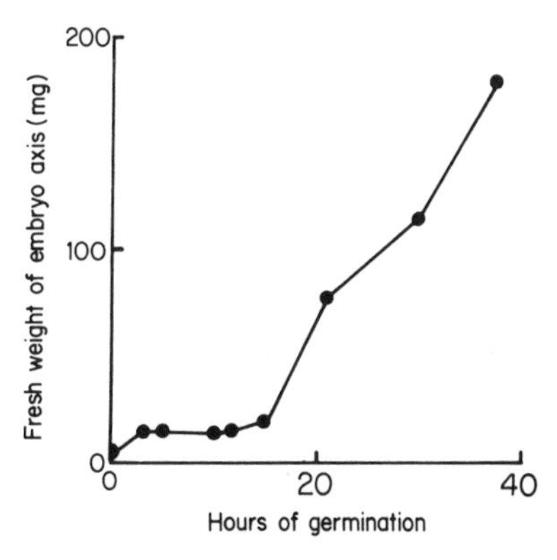

Fig. 3-4 Changes in the fresh weight of the embryo axis in pea seeds germinated at 22°C.

proliferation of subcellular organelles and membrane systems. In the early stages of germination, for example, synthesis of mitochondrial membrane components is largely associated with repair and renewal of membranes, with the consequent restoration of respiratory efficiency. However, later on, significant increases in the amounts of respiratory enzymes occur, and later still, during the cell elongation phase, there is an increase in the number of mitochondria. The endomembrane systems of the cell – the endoplasmic reticulum and the Golgi bodies – also proliferate. The Golgi bodies are active in the transport of polysaccharide material to cell walls. Vacuoles arise from the endoplasmic reticulum, and it is the expansion of these vacuoles (by water uptake), and their subsequent coalescence, which causes the elongation of the cells in the radicle.

In addition to proliferation of organelles and membranes, there are also increases in the amounts of many enzymes. A large proportion of the enzymes are the day-to-day 'working' enzymes of the cell, regulating the vast array of normal metabolic sequences. In addition to these, there is also a significant increase in the activity of those enzymes specifically associated with the renewal of cell division. These include ribonucleotide reductase (a key enzyme in the synthesis of deoxyribonucleotides, the building blocks for DNA synthesis), and DNA polymerase, which synthesizes new DNA molecules.

Increases in the amounts of nucleic acids also occur (Fig. 3-5); the increases in the amounts of ribosomal RNA (and hence of ribosomes), transfer RNA and messenger RNA provide a means of supporting increased rates of protein synthesis (but see Section 3.3.4) and also a preparation for cell division. Replication of DNA within the meristematic cells of the radicle is clearly a specific preparation for cell division. In the pea and its relatives, DNA replication is initiated at about the same time as the start of the large-scale accumulation of RNA (Fig. 3-5), but in many other species, including tomato and radish, DNA replication occurs

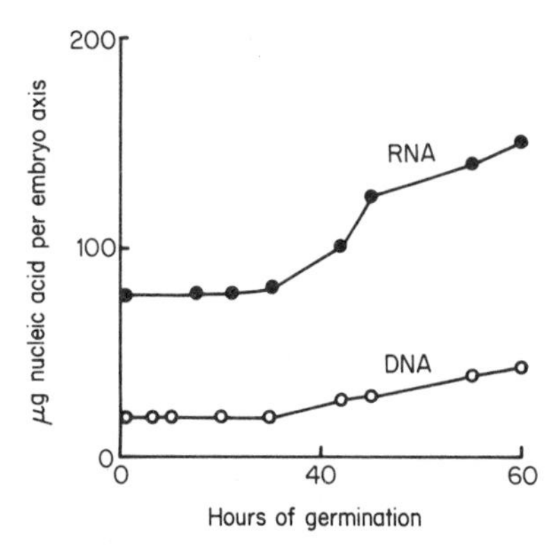

Fig. 3-5 Changes in the nucleic acid content in the embryo axis of pea seeds germinated at 22°C.

somewhat later than the onset of RNA accumulation. Some hours after the initiation of DNA replication, cell division itself is initiated. As has been emphasized already, the early growth of the radicle is achieved by cell elongation without the addition of any new cells. Cell division is thus a relatively late event, occurring well after emergence of the radicle.

3.3.3 Mobilization of reserves

As described in Chapter 1, most seeds contain appreciable reserves of nutrients, which may include, in different species, carbohydrates, proteins, lipids, amino acids, organic phosphate esters and minerals. The function of these reserves is to sustain the seedling until it is firmly established. The bulk of the reserves are usually stored either in the cotyledons, which are destined to senesce relatively early in the life of the plant, or external to the embryo, in the endosperm. Some of the reserves are also laid down in the growing parts of the embryo, to sustain it in the early stages of its growth. Thus, it is found that storage protein disappears from the embryo axis of pea within the first two to three days of germination.

The massive mobilization of the large reserves, laid down in the cotyledons and/or the endosperm, is a relatively late process in germination (Fig. 3-6). The major reserves, at least in terms of bulk, are all polymers; proteins, polysaccharides and lipids. As such, they cannot be transported to the growing embryo axis and must therefore be hydrolysed to yield smaller, readily transportable molecules. A feature of mobilization of food reserves then, is that large increases are observed in the activity of hydrolytic enzymes. Many of these increases in enzyme activity are prevented if the seed is treated with an inhibitor of protein synthesis, such as cycloheximide. This suggests (although it does not prove) that the increase in enzyme activity is mediated by synthesis of new enzyme protein. However, for a number of these enzymes, the increase in activity is not prevented by treatment with inhibitors. Such enzymes, which include amylopectin-

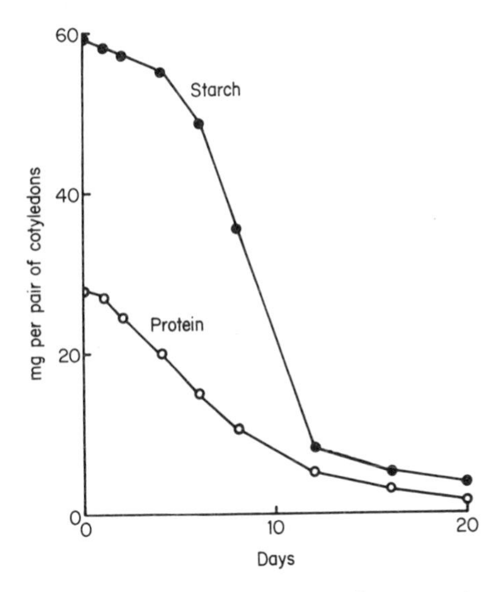

Fig. 3-6 Changes in the starch and storage protein content in cotyledons of pea seeds germinated and grown at 22°C.

1-6-glucosidase (a starch-degrading enzyme) and ribonuclease (which hydrolyses RNA) in pea, and β-amylase (another starch-degrading enzyme) in wheat, are present in an inactive form in the dry seed, and are in some way activated during germination.

3.3.4 Gene expression: new and old messenger RNA

The immediate resumption of metabolic activity which occurs on hydration has already been mentioned, and it is now clear that this resumption includes the synthesis of both RNA and protein. The relationship between the timing of the onset of RNA synthesis and the timing of the onset of protein synthesis was for some years a matter of controversy. Protein synthesis takes place on ribosomes, involves the use of transfer RNA as adaptor molecules, and is directed by messenger RNA (Fig. 3-7). In the 1960s it was widely assumed that messenger RNA was unstable. It therefore came as a surprise when a number of investigators claimed that protein synthesis in seeds was re-initiated early after the start of imbibition: too early, it was argued, for there to be time to make new messenger RNA molecules. It was further claimed that this early protein synthesis was not inhibited by inhibitors of RNA synthesis, again suggesting that synthesis of new messenger RNA was not needed. The implication of these claims was that populations of messenger RNA molecules survived desiccation and could be used to direct protein synthesis during germination. This idea was strengthened by the finding that in developing cotton seeds, induced to germinate precociously, two hydrolytic enzymes, protease (which breaks down protein) and isocitrate lyase

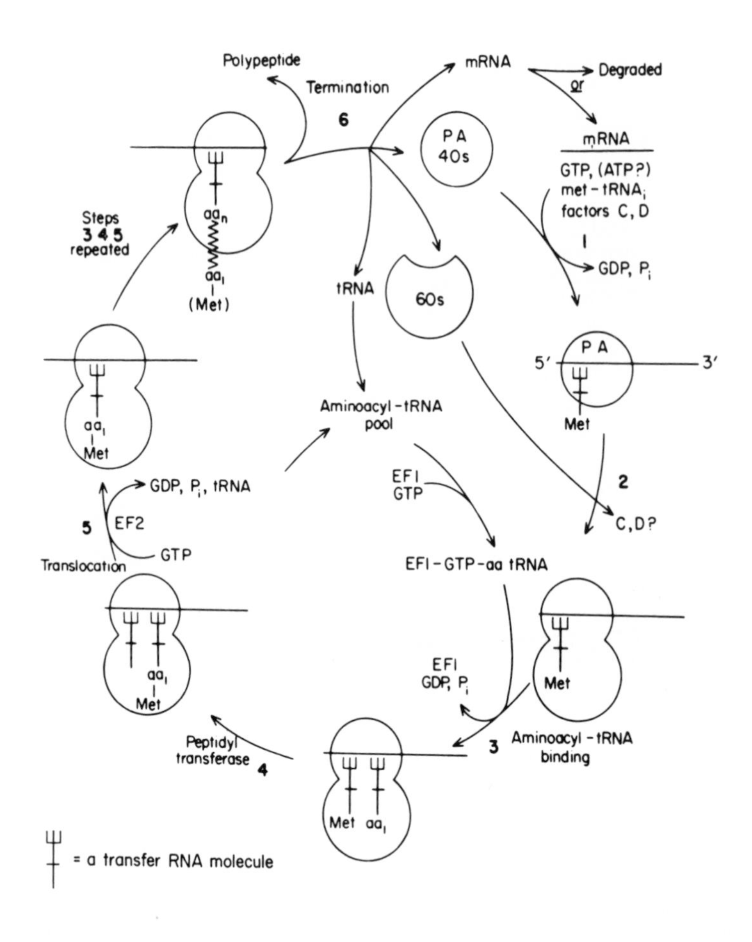

Fig. 3-7 Protein synthesis on plant 80*S* ribosomes. tRNA = transfer RNA; met-tRNA = methionyl tRNA carrying a methionine molecule; aa_l = the amino acid inserted immediately after the initiating methionine molecule; aa_n = the most recently inserted amino acid; P,A = the peptide and amino acid binding sites on the ribosome; 60*S* and 40*S*, the two subunits of 80*S* ribosome; EF1,2 = protein factors necessary for elongation of the polypeptide chain; Factors C,D = factors necessary for binding of mRNA to the 40*S* subunit; GTP = guanosine triphosphate (an energy carrying molecule similar to ATP); GDP = guanosine diphosphate; P_i = inorganic phosphate (i.e. a phosphate ion).

(involved in lipid mobilization) could apparently be synthesized without the need for RNA synthesis (see Chapter 1).

This interpretation of the data did, however, suffer from a number of defects, the most important of which was that the experimental techniques available at the time did not permit the isolation and unequivocal identification of messenger RNA. The development, in the early 1970s, of techniques for isolation of messenger RNA (Fig. 3-8) has, however, led to a great clarification of the situation regarding messenger RNA in seeds. The isolated messenger RNA can then be used to direct protein synthesis on ribosomes in the test tube. When such experiments were first carried out with seeds, it became clear that unimbibed seeds did indeed contain messenger RNA. This is now known as long-lived or stored messenger RNA, and is known to occur in a number of other stages of plant development in addition to germination. Furthermore, it is now generally accepted that messenger RNA molecules are rather more stable than was formerly supposed.

What, then, is the function of the long-lived messenger RNA in seeds? Inclusion of radioactively labelled amino acids in protein synthesis experiments conducted in the test tube leads to the formation of labelled proteins. Thus, the proteins coded for by any particular messenger RNA population may be detected and compared with known proteins. Such experiments lead to the conclusion that a high proportion of the messenger RNA in an ungerminated seed is 'left over' from seed development. In pea,

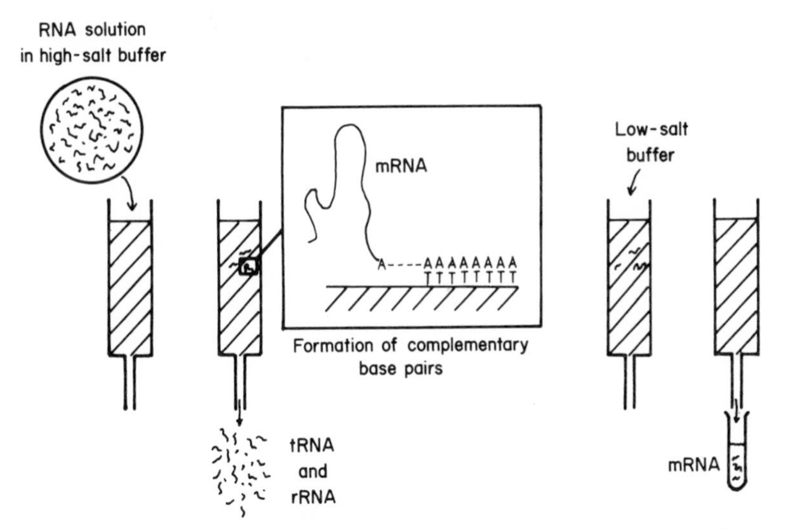

Fig. 3-8 Isolation of messenger RNA. The RNA preparation is applied to a column of oligo-dT-cellulose. The poly-A tails of the messenger RNA molecules hydrogen bond to the oligo-dT (i.e. form complementary base pairs) and the remainder of the RNA passes through. The messenger RNA can then be washed from the column with a buffer of very low ionic strength (to suppress the hydrogen bonding).

oil-seed rape and cotton, for example, long-lived messenger RNA supports the synthesis, in the test tube, of a range of proteins, including the seed storage proteins. This conclusion is supported by the fact that the amount of messenger RNA actually declines in early germination, presumably as 'left-over' messages are degraded, (prior to the increase in messenger RNA content which is associated with renewed growth). However, about 40% of the long-lived messenger RNA molecules are actually used to direct protein synthesis in the germinating embryo. Are these then messenger RNA molecules which code for proteins which are specific to or essential for early germination? Perhaps disappointingly, the answer seems to be, No! The *long-lived* messenger RNA molecules which are used in early germination appear to be no different from those which are *newly synthesized* in early germination. It is therefore probable that both the long-lived messenger RNAs and the messenger RNAs synthesized in early germination code for the general 'working' proteins of the cell, such as the enzymes involved in intermediary metabolism. Later on, as the embryo resumes active growth, messenger RNAs for new proteins, including those specifically involved with cell division, are synthesized. Curiously, two enzymes in cotton seeds, protease and isocitrate lyase, which are probably coded for by long-lived messenger RNA, are involved in the hydrolysis of food reserves. The lateness of this event (see Section 3.3.3) makes it difficult to understand why the two enzymes should be coded for by a messenger RNA which is present in the dry seed.

3.4 Plant growth substances in germination

The involvement of plant growth substances (plant hormones) in the prevention of precocious germination and in the breakage of dormancy has already been dealt with in Chapter 1 and 2. There is, however, much less information concerning the role of growth substances in germination itself. Although growth substances have been demonstrated to be involved in the regulation and co-ordination of plant growth at a variety of developmental stages, there is no clear evidence at all that they are involved in controlling the growth of the embryo during germination. The exogenous application of growth substances to embryos certainly affects their growth (abscisic acid, for example, strongly inhibits embryo growth) but this does not prove that growth substances are normally involved.

The only convincing evidence for the involvement of growth substances in germination concerns the mobilization of food reserves in cereal grains, of which barley may be taken as typical. In barley, as in other cereals, the bulk of the reserves are stored external to the embryo, in the endosperm (which consists of dead cells in cereals and other grasses). Hydrolytic enzymes are secreted from a layer of living cells, the aleurone layer, situated around the endosperm, resulting in the breakdown of the stored polymers to small molecules. These are taken up via transfer cells by the scutellum (cotyledon) and transported to the growing parts of the embryo.

Barley grains may be cut into two halves, relatively easily, one containing the embryo and the other containing the bulk of the endosperm. If both half-grains are 'germinated', the half without the embryo does not exhibit any increase in the activity of the hydrolytic enzymes, such as α-amylase. If the embryo-less half-grains are 'germinated' in the presence of gibberellic acid, then normal increases in enzyme activity occur, and hydrolysis of the reserves proceeds as usual. In a normal grain, gibberellic acid is secreted by the embryo and stimulates the production of hydrolytic enzymes in the aleurone layer. The enzymes are then secreted into the endosperm (Fig. 3-9).

Unlike that in barley, the endosperm in most other endospermous seeds consists of living cells, and in such seeds, there is no clear evidence at all that the embryo controls the breakdown of the endosperm reserves via secretion of growth substances. Finally, for those seeds in which the reserves are laid down in a part of the embryo, namely the cotyledons, it is becoming

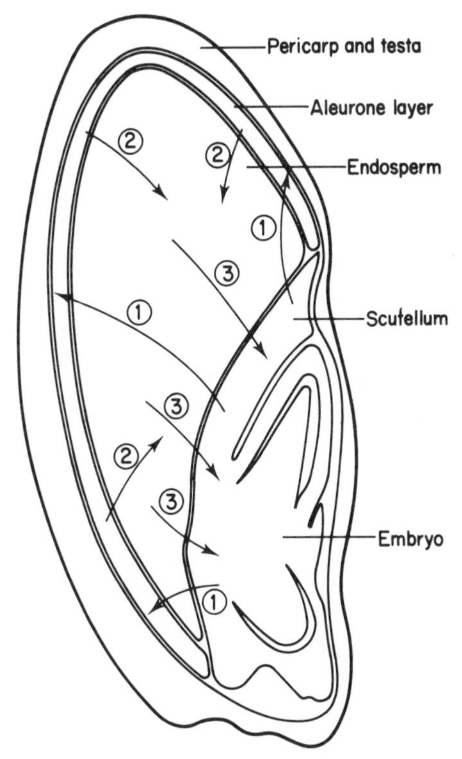

Fig. 3-9 Diagram illustrating the control of the mobilization of reserves in germinating barley seeds. **(1)** The embryo secretes gibberellic acid (GA) to the aleurone layer. **(2)** In response to GA the aleurone cells synthesize hydrolytic enzymes and secrete them into the endosperm. **(3)** The hydrolysis products are taken up by the embryo.

apparent that the growing part of the embryo, the embryo axis, does not have any hormonal influence at all on the hydrolysis of reserves. So, for example, in cotyledons of pea from which the axis is removed, normal increases in enzyme activity occur. However, removal of the axis does prevent hydrolysis of the reserves (despite the presence of the hydrolytic enzymes). The reason for this is not that growth substances from the embryo axis are needed. Rather, it is a simple mass-action feedback effect: in the absence of a 'sink' (i.e. in the absence of a growing embryo axis) to receive the products of hydrolysis, extensive hydrolysis of the reserves is inhibited.

4 Seed Physiology and Ecology

4.1 The seed as a dispersal unit

The seed is the culmination of a phase of plant development which starts with fertilization and finishes with desiccation. In the desiccated state the seed is able to withstand a variety of environmental stresses without undue damage and this makes the seed an ideal unit of dispersal. In many species there are specific seed dispersal mechanisms. The explosive cracking of the pods of gorse (*Ulex*), a familiar sound on warm summer days in heathland areas, propels the seeds away from the parent bush. In species of cranesbill (*Geranium*) and balsam (*Impatiens*) the seeds are propelled by a spring-like mechanism, caused by differences in the drying rate in different parts of the fruit. For seeds enclosed in fleshy fruits, dispersal occurs after ingestion of the fruits by birds and animals followed by passage of the seeds through the alimentary system. In a small number of species, passage through the alimentary system of a bird or mammal is actually needed to break dormancy. Tomato seeds in the fruit for example, are coated by a mucilage which contains germination inhibitors. Passage through a mammalian alimentary system removes this inhibitor, as is vividly demonstrated by the enormous number of tomato plants which appear at sewage treatment works!

A particularly interesting example is seen in seeds of *Calvaria major*, a tree which grows on the tropical island of Mauritius. All trees of this species on the island are over 300 years old, and it is clear that no seeds have germinated for this length of time, despite the fact that the few remaining trees fruit regularly. It appears that the fruit, a drupe (rather like a plum) was eaten by the dodo, and that passage through the alimentary system of this bird softened the fruit stone and so allowed the enclosed seed to germinate. With the extinction of the dodo, there was no mechanism to allow seed germination. Recent experiments have shown that ingestion of the fruits of *C. major* by turkeys has the appropriate effect, and so germination of the seeds has now been achieved for the first time in over 300 years.

Other seed dispersal mechanisms include structures on the enclosing fruit for catching on the coats of animals, as in cocklebur (*Xanthium*) and structures enabling the seed to be carried by the wind, such as the pappus on dandelion (*Taraxacum officinale*) seeds or the winged fruits of sycamore (*Acer pseudoplatanus*) and ash (*Fraxinus excelsior*). Seeds of plants growing in or near water often have flotation mechanisms, enabling the seed to be carried away by water. A particularly good example is the fibrous husk of the coconut, which supports the seed in the sea, enabling it to be dispersed from island to island.

Even in species which lack specific dispersal mechanisms, seeds may be dispersed by wind (as with very small seeds), by rolling, by movement in soil and/or surface water or even by being collected and then buried by animals. Squirrels, for example, are very efficient, if unwitting, dispersal agents for seeds of several forest trees.

In addition to seed dispersal mechanisms, particular facets of seed physiology, such as germination requirements, dormancy (or lack of dormancy) and particular requirements for breakage of dormancy all serve to maximize the chances of a seed being dispersed to, and then exploiting, its particular habitat. The remaining sections of this chapter deal with these ecological aspects of seed physiology in more detail.

4.2 Germination and climate

Anyone who has grown tomatoes, peppers or aubergines (egg plants) from seed will know that the seeds need to be kept at 20–25°C for maximal germination. Below 10°C tomato seeds will not germinate at all, but failure to germinate at this low temperature must not be equated with dormancy. The seeds are not dormant, and no amount of dormancy breaking treatment (of any kind) will induce the seeds to germinate at, for example, 8°C. At the other end of the scale, these seeds will germinate at temperatures up to 40°C (Fig. 4-1). By contrast, lettuce seeds will germinate at temperatures well below 10°C, although germination may be slow at these lower temperatures. The upper limit for germination is also lower: 25–30°C (Fig. 4-1).

The relationship between temperature and germination behaviour is related to the temperatures normally experienced by the plants, during their growing season. However, in some species, the relationship between temperature and germination is not quite so straightforward. In campion

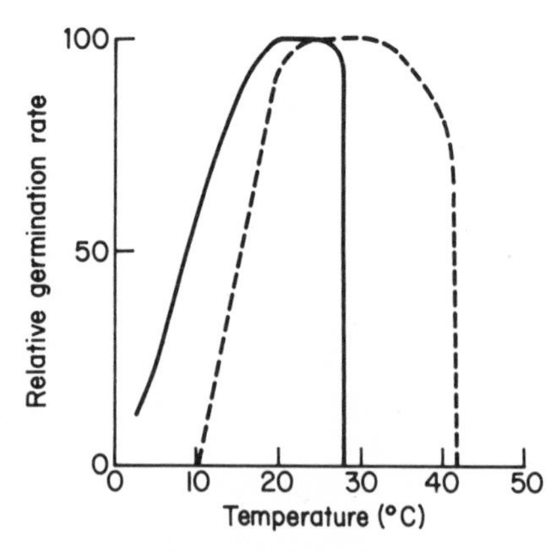

Fig. 4-1 Effects of temperature on the germination of lettuce (——) and tomato (-----) seeds.

(*Silene vulgaris*) and in other plants with a wide distribution in Western Europe, germination behaviour is related to the prevailing temperature in the non-growing season, i.e. winter. Seeds collected from the Mediterranean region have *lower* temperature minima and maxima for germination than seeds collected from Central Europe, where the winters are much colder. The lower the temperature in the unfavourable season, the higher the temperature required for germination. This is best seen as a mechanism for preventing germination should an unseasonably mild spell of weather occur before winter is really over.

4.3 Dormancy and seed ecology

In the previous section, germination behaviour was seen to be related both to exploiting suitable growth conditions for the plant and to avoiding unfavourable conditions. Both these aspects are also seen in consideration of seed dormancy and dormancy-breaking mechanisms.

4.3.1 Seeds with a light requirement

The requirement for light to break dormancy is an obvious adaptation enabling the seed to germinate in open habitats. A seed lying beneath a leaf canopy is effectively in the dark, at least as far as phytochrome is concerned, since passage of light through leaves gives a range of wavelengths which causes conversion of Pfr to Pr and hence leads to seed dormancy (see Chapter 2). Similarly, there is a relationship between the chlorophyll content of the tissues enclosing the embryo (seed coat plus maternal tissues) and a light-requirement for breakage of dormancy. In a survey of a wide range of species, chlorophyll contents greater than 0.6 mg g^{-1} dry weight in the surrounding tissues were correlated with seed dormancy; seeds which matured within less green surrounding tissues did

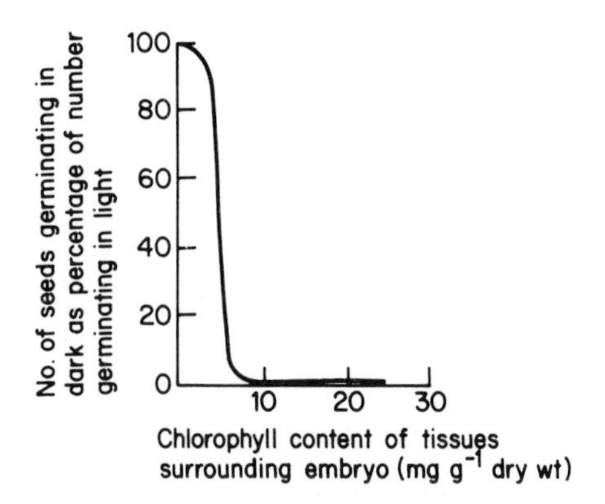

Fig. 4-2 Relationship between chlorophyll content of tissues surrounding the developing embryo and the requirement for light exhibited by the mature seed. Based on data from 18 different species.

not show a light requirement (Fig. 4-2). This is another indication that light passing through green tissues is enhanced for far-red wavelengths and that this leads to the establishment of a high Pr/Pfr ratio in the maturing embryo.

Even in the absence of a canopy, light-requiring seeds buried deep in the soil will not germinate, but if the soil is disturbed in any way, seeds may be brought to the surface and then germination will occur. Thus, many of the species in which the seeds need light to break dormancy are species which need open ground for seedling establishment. Not surprisingly, many of these are common weeds, exploiting habitats such as newly dug garden soil!

4.3.2 Seeds requiring alternating temperatures

Although a light-requirement is one major factor controlling the germination of seeds in open habitats, it is not the only one. Indeed, a significant proportion of the seeds which germinate in open ground do not require light, but need alternating temperatures to break their dormancy. Throughout a 24 hour period, there are fluctuations in soil temperature, but these fluctuations are very much damped out by vegetation cover and by increasing depth in the soil (Fig. 4-3). Further, the maximum and minimum temperatures vary through the year, as does the magnitude of the fluctuation. These factors are also affected by the size of the gaps in the cover. As was described in Chapter 2, those seeds with a requirement for alternating temperatures exhibit a great variety in their needs in respect of the magnitude of the various parameters involved. Figure 4-3 illustrates the relationship between gaps in the vegetation cover, soil depth, the

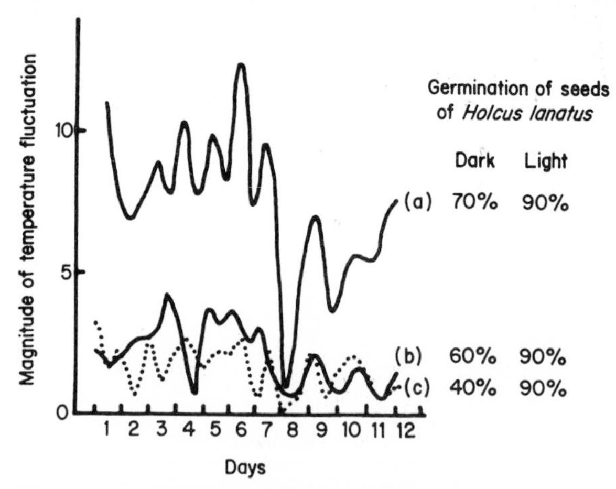

Fig. 4-3 Diurnal temperature fluctuations in soil in late summer in Northern England. **(a)** Open ground, 10 mm depth. **(b)** Beneath cover of pasture grasses, 10 mm depth. **(c)** Open ground, 50 mm depth. Alongside the graphs the predicted germination percentages for seeds of *Holcus lanatus* (Yorkshire Fog) are presented, based on the observed germination behaviour of seeds in the laboratory.

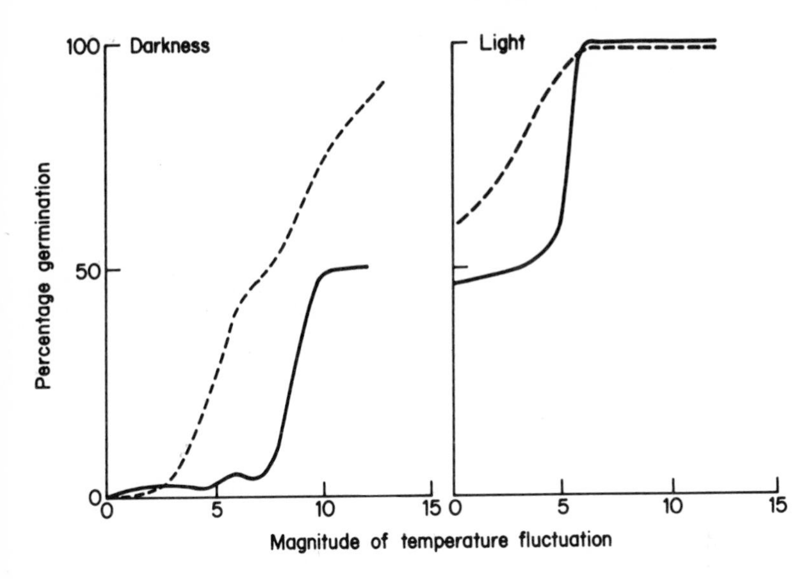

Fig. 4-4 Germination of *Rumex obtusifolius*, broad-leaved dock (——) and *Deschampsia cespitosa*, tufted hair-grass (------), **(left)** in darkness and **(right)** in light, in response to diurnal temperature fluctuation.

resulting diurnal temperature fluctuation and the germination of one grass species, *Holcus lanatus*. Figure 4-4 illustrates two different requirements in respect of the magnitude of the temperature fluctuation, in two different species which colonize open ground. The specific requirements of seeds of a given species thus help to promote their germination in open spaces at appropriate times of year.

Finally, it should be noted that there are many species in which the effects of light and fluctuating temperatures are additive. One such is the grass, cocksfoot (*Dactylis glomerata*). This feature of their physiology clearly maximizes the chances of establishment in an open habitat, since as the openness of the habitat increases so does the proportion of seeds which germinate.

4.3.3 Seeds with a requirement for low temperature

The need for a period in the cold in order to break dormancy is a mechanism to avoid exposure of seedlings to potentially damaging low temperatures. The phenomenon is thus widespread in temperate, cool-temperate, sub-arctic and montane regions. The British flora, for example, contains many species with cold-requiring seeds, a small selection of which is shown in Table 4.1.

Since a cold requirement is a mechanism for avoiding germination before or during the winter, it might be expected that the length of cold treatment required will vary according to the length and severity of the winter season

Table 4.1 Some of the many British plant species with seeds which require a cold treatment to break dormancy.

Caltha palustris	Kingcup, Marsh marigold
Viola arvensis	Field pansy
Oxalis acetosella	Wood sorrel
Impatiens parviflora	Small balsam
Acer pseudoplatanus	Sycamore
Prunus spinosa	Sloe, Blackthorn
Heracleum sphondylium	Hogweed
Mercurialis perennis	Dog's mercury
Corylus avellana	Hazel
Fraxinus excelsior	Ash
Solanum dulcamara	Woody nightshade
Lonicera periclymenum	Honeysuckle
Allium ursinum	Ramsons, Wild garlic

normally experienced by the plant. Examination of a range of species illustrates that this is indeed so. For example, the Norway maple (*Acer platanoides*), a native of Northern Europe, is widely planted as an ornamental or parkland tree in the British Isles. Although the trees produce copious fertile seeds, very few of these germinate naturally, particularly in southern and south-western Britain. If, however, the seeds are collected and kept in the cold room at 2–4°C, for three to four months, the seeds germinate readily (Fig. 4-5). A closely related species, sycamore (*Acer pseudoplatanus*) has a more southern and western distribution. Its seeds require a rather shorter period of exposure to the cold than those of Norway maple (Fig. 4-5) and hence germinate naturally even in the milder parts of the British Isles. Similar variations in depth of dormancy may even be seen within one species if that species has a wide distribution range. One such species is ash (*Fraxinus exelsior*); seeds from trees in the northern part of the range require a longer exposure to low temperatures than seeds from trees in the southern and south-western parts of the range.

4.3.4 Seeds with a requirement for warmth

Seeds which have a requirement for warmth to break dormancy are not as common, at least in the Northern hemisphere, as those which need low temperatures. Typically, a requirement for warmth is shown by seeds of spring-flowering plants. The seeds are shed in late spring or early summer and are exposed to summer temperatures. They germinate in the autumn and the young seedlings overwinter to resume growth in the spring. Germination thus does not occur in high summer, when the seedlings may be subjected to hot or dry conditions. The autumn germination means that the seedlings can become established before potential competitors or before the leaf canopy thickens. For species showing this type of strategy, it is obviously essential that the young seedlings are able to withstand the winter weather conditions, either because the winters are mild, as in Southern Europe, or because the seedlings are frost-hardy, or because, as in bluebell (*Hyacinthoides non-scripta*), the seedling is somewhat insulated from the weather under the ground layer of leaf litter.

It should be noted that not all spring-flowering plants have seeds which germinate in late summer. In many, the seeds are cold-requiring and in some, such as hawthorn (*Crataegus*) the seeds require a period of warmth followed by a cold period. These responses thus 'record' the conditions experienced by the seeds and ensure that germination is delayed until spring.

4.3.5 Seeds with a requirement for fire

The extreme in a requirement for warmth is seen in seeds which require fire in order to germinate. As might be expected, species with such seeds are typically plants of habitats which are prone to fire. Indeed, many such species are actually dominant in such habitats. They include various bush and tree species of the family Leguminoseae, which grow in sub-tropical scrub-land, and also those shrub species of the family Ericaceae which grow on heathland (in the latter, the fire requirement is by no means absolute: see Chapter 2).

The requirement for a brief very high temperature treatment (combined in the heathers with 'seed-banking': see Section 4) means that these species are readily able to recolonize open spaces caused by fire and are obviously well able to replace any mature plants destroyed by the fire. However, it should not be assumed that all plants which colonize fire sites have seeds which need fire. Many are simply colonizers of open ground, with seeds

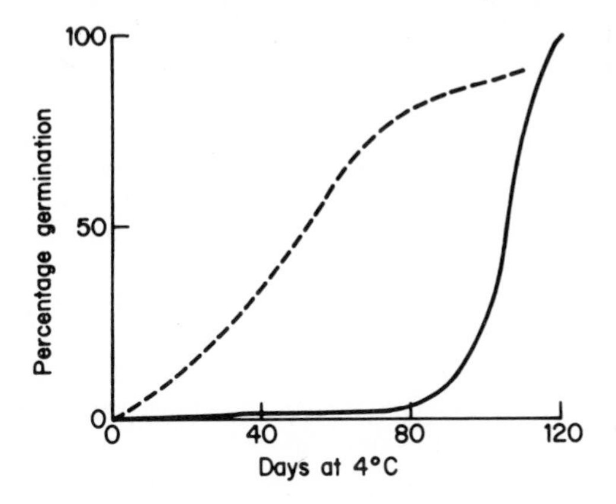

Fig. 4-5 Breakage of dormancy by cold treatment in two species of *Acer. A. platanoides*, Norway maple (——) and *A. pseudoplatanus*, Sycamore (------). Germination trials were carried out at 17–20°C.

which require light or fluctuating temperature to break dormancy (as discussed in Section 4.3.1 and 4.3.2) whilst for others, the fire acts to bring the soil temperature into the optimum range for germination (Section 4.2).

4.3.6 Desert ephemerals

Seeds need water for germination, but ungerminated seeds of many species can survive for many years in the dry state (see Chapter 5) and will germinate when eventually provided with water. Many readers will be familiar, at least from pictures, with the spectacular displays of colour afforded by blooms of desert ephemerals soon after a heavy rainstorm. In desert regions, significant rainfall may not occur for several years and the seeds of these ephemerals survive in the quiescent state during the dry periods. The seed coats contain inhibitors of germination which are washed out by rain, thus allowing the seeds to germinate. However, in order for the plants to complete their life-cycle, albeit brief, and to set seed, it is likely that more than a passing shower is required. In these species there is thus a danger that a succession of short showers, even if separated by months or years, could wash out all the germination inhibitor whilst not giving enough water for successful completion of the life-cycle. This danger is avoided by an interesting mechanism. If the rain stops before all the inhibitor is washed away, the partially hydrated seed synthesizes more inhibitor. So, the seed does not germinate in response to a short shower, or to a drawn-out series of short showers, and is able to survive the partial hydration and then slow dehydration while awaiting enough rain to wash out all of the inhibitor.

4.4 Non-dormant seeds

Both in this chapter and in Chapter 2, the widespread occurrence of seed dormancy, particularly in wild plants, has been stressed. However, there are also many wild plants which do not have dormant seeds, or in which a proportion of the seeds are not dormant. In a survey of 403 British species of wild plants, 32% of species had seed populations in which 50% or more of the seeds were not dormant. Non-dormant seeds can germinate as soon as they are shed, provided of course that they are rehydrated at appropriate temperatures and oxygen tensions. In general such plants occupy habitats where there are likely to be patches of open ground, such as arable land or well grazed downland. Typical plants in this category are the ruderals, short-lived weeds which readily colonize bare ground, such as Bitter Cress (*Cardamine hirsuta*) and annual grasses such as *Poa annua* and *Aira praecox*. These annual grasses are in fact winter annuals, and the lack of dormancy means that germination occurs before the winter sets in, and that the seedling has a start over its competitors in the following spring. In some ways this strategy resembles that of plants whose seeds pass the summer in the dormant state (Section 4.3.5) except that in these annual grasses there is no mechanism to avoid exposure of the seedling to the possibly damaging heat of summer.

4.5 Polymorphism

It was noted in Chapter 2 that beet seeds are particularly frustrating for the grower because different seeds exhibit different depths of dormancy. This phenomenon is known as polymorphism and is widespread in wild plant species (as noted in the preceding section and in Chapter 2). The control of this polymorphism is not understood, although it is clear that the position of the ripening seed on the parent plant or within the flowering head can affect depth of dormancy, as is seen in many members of the Compositae (daisy family). The most spectacular example of this is seen in cocklebur (*Xanthium pennsylvanicum*) where pairs of seeds occur together in a burr (which is dispersed by hooking into the fur of passing mammals). In the burr, the upper seed is deeply dormant, whilst the lower seed is hardly dormant at all. Whatever controls polymorphism, the adaptive significance of the phenomenon is clear: a range of differing depths of dormancy (including non-dormancy) within a seed population ensures that germination of that seed population is spread out over a period of months, or even years, hence maximizing the chances of establishment in a suitable habitat.

4.6 Seed banks

The strategy of spreading germination over a period of time is seen particularly in those plants which form seed banks. A typical cubic metre of soil, whether in a natural or cultivated habitat, contains many thousands of seeds. This may be illustrated by digging a soil sample to a depth, for example, of 15 cm, spreading the soil out in a seed tray and keeping it moist in the light. Many seedlings will appear within a few days, and even more will appear if the temperature is allowed to (or made to) fluctuate up and down with a 24-hour periodicity. The range of species which appear in such experiments, of course, varies according to the source of the soil sample, but it is clear that any soil will contain an array of seeds, from a wide variety of species, waiting to germinate. A small selection of species which form seed banks is shown in Table 4.2. It will be noted that they are all species which shed large number of small seeds. This means that many of the seeds are likely to penetrate the soil via cracks and channels and thus become vertically distributed within the topmost few centimetres of the soil profile.

As might be expected, many of the species which establish extensive seed banks have seeds which are dormant (or at the least, have seed populations composed partly of dormant seeds). The dormancy-breaking requirements include light, as in thyme (*Thymus praecox*), fluctuating temperatures, as in chickweed (*Stellaria media*) or long-term damp storage in the soil to induce seed coat deterioration as in heather (*Calluna vulgaris*). In some instances, as noted earlier, this type of seed coat dormancy may also be broken by fire.

However, not all seed-banking species shed dormant seeds. In those species where the seeds are not dormant, some seeds may germinate soon after shedding. Other seeds will penetrate the soil to depths where the

Table 4.2 A selection of the British plant species which form persistent seed banks.

Species		*Mean seed weight (mg)*
Arabidopsis thaliana	Thale cress	0.02
Stellaria media	Chickweed	0.35
Sagina procumbens	Pearlwort	0.02
Saxifraga tridactylites	Rue-leaved saxifrage	0.01
Epilobium hirsutum	Hairy willow-herb	0.05
Polygonum aviculare	Knotgrass	1.45
Digitalis purpurea	Foxglove	0.07
Calluna vulgaris	Ling, heather	0.03
Origanum vulgare	Marjoram	0.10
Thymus praecox	Wild thyme	0.11

conditions do not favour germination, perhaps because of low temperatures or poor aeration and remain in the ungerminated state (although not dormant) until conditions improve. Such improvement may arise by opening of the vegetation cover, by soil disturbance, by increases in the ambient temperature or by changes in soil water potential.

Finally, it seems very likely that a range of seed-banking species exhibit the phenomenon of secondary dormancy (see Chapter 2). This means that if the seeds are not exposed to suitable germination conditions they become dormant and then need a specific dormancy breaking factor to induce germination. This has been observed in species of willow-herb (*Epilobium*) and dock (*Rumex*). Although detailed surveys of secondary dormancy in relation to seed-banking have not been carried out, it seems quite probable that this particular pattern of seed behaviour is quite widespread in seed banking species.

Species which form seeds banks thus produce large numbers of long-lived seeds. These features, particularly when combined with polymorphism with respect to dormancy, or with the development of secondary dormancy, serve to spread the germination of a particular seed batch over a long period of time. Again, this illustrates the importance of characteristics of seed physiology as adaptive mechanisms leading to the maximal exploitation of suitable habitats.

5 Seed Physiology and Agriculture

5.1 Seeds in agriculture

Seeds fulfil two very important roles in agriculture and horticulture. Firstly, the seeds serve as the basic propagule, the starting point, for many crops. Many agriculturally important plants are annuals, including cereals (wheat and barley) and legumes (peas and beans), or biennials harvested in their first year, such as carrot and beet. The great majority of these are propagated from seed and so the grower requires regular supplies of seed which germinate well. It is therefore relevant to consider how many seeds in a given batch will germinate and how long they will take to germinate under field conditions.

Secondly, seeds may also represent the end product, the crop, which the grower will harvest. These include the vast variety of seeds used directly in human and animal nutrition, particularly the world's major cereal crops - barley, maize (corn), wheat and rice - and numerous legumes. Also included here are those seeds which are used almost exclusively to yield compounds such as lipids which may be used for nutritional or other purposes. Seeds in this category include soybean, sunflower, oil-seed rape, castor bean and mustard. For use of seeds as the final harvest, the grower requires reliable embryogenesis and seed set under field conditions to give a crop of high yield and quality.

5.2 Germination in agriculture

5.2.1 Seed longevity, viability and vigour

It is not normal agricultural or commercial practice to store seeds for long periods before use. Indeed, seeds collected at the end of a given summer are usually sown the next spring. Nevertheless, the ability of seeds to survive storage is important and interesting, and is relevant in situations where crop failure (because, for example, of drought or flooding) may lead to the use of seeds stored for a longer time than usual.

In order to examine seed longevity, two other terms must be introduced: seed viability and seed vigour. If 100 seeds are set to germinate and 99 of them do so, then the seed batch is obviously a high viability batch; if 50 germinate, the batch may be termed a median viability batch, and if none germinate, the batch is non-viable. Seed vigour is related to the speed of germination. Seeds of high vigour germinate rapidly; seeds of low vigour germinate more slowly. The difference between viability and vigour is illustrated in Table 5.1. Both batches of seed are of high viability, since over 80% germinate. However, in one batch, the seeds are much less vigorous than the other, this being reflected in the difference in the T_{50} values (time

Table 5.1 Germination of high and low vigour carrot seed batches at two different temperatures.

Seed batch		*Germination temperature* *10°C*	*25°C*
Low vigour	% germination	82.5	85.3
	T_{50}* days	12.1	2.8
High vigour	% germination	81.2	84.5
	T_{50}* days	5.2	2.3

*Time taken for 50% of the seeds to germinate

taken for 50% of the seeds to germinate) particularly at 10°C.

The data in Table 5.1 also illustrates another important aspect of seed vigour, namely that low vigour may only be exhibited under non-ideal germination conditions. Whereas there was a marked difference in T_{50} between the two seed batches in a germination trial at 10°C, the batches had very similar germination characteristics at 25°C. This relationship between the manifestation of low vigour and environmental conditions is important for the grower. A laboratory trial of seeds carried out at 20–25°C with a controlled water supply does not indicate how a seed batch will perform in the field, where the soil temperature in temperate and cool-temperate regions may well be around 5°C when the seeds are sown, and where water supply is not closely controlled.

The relationship between seed longevity, vigour and viability may be envisaged as follows. During seed storage, changes take place within the seed which are detrimental to its ability to germinate. The nature of these changes will be discussed later. As the extent of this deterioration increases so individual seeds lose vigour and germinate more slowly. This will be reflected in a higher T_{50} for the batch of seeds; initially this may occur only under non-ideal conditions, but eventually will also be seen under ideal conditions. Continuation of the damaging changes within the seeds eventually results in individual seeds being unable to germinate; in other words they become non-viable. Further storage will lead to more and more seeds becoming non-viable until the whole seed batch fails to germinate.

This facet of seed physiology is so important for the grower that it has received much attention from research workers. Efforts to identify a single primary cause of loss of vigour and viability have failed, probably because there is no individual lesion which governs the overall process. Rather, it appears that changes in germinability occurring during storage reflect a wide range of lesions in seed function. Some of the lesions are more dramatic versions of the damage that often occurs as a result of desiccation and rehydration (Chapter 3). Ribosomal RNA molecules may become partially degraded, DNA molecules are broken, eventually leading to chromosome breakage and membranes become more and more leaky. Other detrimental changes which occur include breakdown of messenger and transfer RNA. As seen in Chapter 3, a limited amount of damage may be repaired during the early stages of germination. As the extent of the damage becomes greater, so it takes longer to repair and hence germination takes longer, i.e. seed vigour declines. Eventually the damage becomes too extensive to repair and the seed is then non-viable.

What then is the cause of this damage? The answer to this question is complex, but there is a unifying factor in the complexity and that is water. Water is a ubiquitous component of living cells, existing partly in ionized form. Many biological macromolecules and macromolecular complexes such as membranes depend for their conformation on a water shell or on an aqueous environment which allows correct disposition of the hydrophilic (water-loving) and hydrophobic (water-repelling) regions of the molecule or molecular complex. Desiccation upsets this balance between the macromolecules and their aqueous environment, and rehydration during germination may actually cause damage whilst the normal equilibrium is being re-established. This appears to be a normal part of germination (Chapter 3) and probably does not affect the vigour of the seed.

However, water is not chemically inert; pure water has an effective concentration of about 55M and is a very efficient agent of hydrolysis. Many of the important chemical linkages in biological molecules are prone to hydrolysis, for example phosphate esters (as in ATP), phosphodiester linkages (as in nucleic acids), peptide linkages (as in proteins) and glycosidic linkages (as in oligo- and polysaccharides), and random degradation is a grave danger in an aqueous environment. Indeed, some biological molecules such as ATP are hydrolysed at a measurable rate in aqueous solutions in the test tube. In the normal cellular environment much of this random hydrolysis is prevented by the ordered arrangement of the macromolecules and also because many of the compounds which are subject to hydrolysis are actively engaged in cellular metabolism. As such, they are complexed with enzymes which allow access of water to vulnerable linkages only under the strictly regulated conditions necessary for the reaction being catalysed. Nevertheless, in conditions where there is a rapid rehydration, as in germination, some random hydrolysis is bound to occur, particularly in the period prior to re-establishment of the usual ordered arrangements within the cell. Again, this may be regarded as a normal feature of germination, and unless it is unusually excessive, will not affect seed vigour to a measureable extent.

These 'normal' deleterious processes occurring more or less routinely in the life of the seed help in understanding the long-term deterioration which occurs during seed storage. Even at the very low moisture content of seeds, water can act as a hydrolytic agent. Although its low concentration in the seed reduces the chances of random hydrolysis, the lack of the usual order within the cell to some extent counteracts this. Given enough time then, random hydrolysis will occur. Further, some of this hydrolysis may actually be enzyme-mediated. Although enzymes work at a rate so slow as to be immeasurable in dry seeds, the lack of the normal ordered arrangements of macromolecules and complexes may allow hydrolytic enzymes access to substrates, and given long enough there is a statistical probability that some hydrolysis will occur.

Water, then, has an important role to play in these random hydrolytic processes, and within the normal range of moisture contents exhibited by seeds, the higher the moisture content, the greater is the damage suffered. Further, since the damage is essentially chemical or biochemical, temperature also affects the rate at which damage occurs. So, seeds of low moisture content stored at low temperature suffer much less damage than seeds of high moisture content stored at high temperature. A number of examples from current commercial or agricultural practice will illustrate this. Firstly, when seeds of plants such as carrot or turnip are collected for sale to growers, they are collected from areas where the seed is exposed to warm, dry conditions during maturation, thus ensuring as low a moisture content as possible at harvest. Carrot seed for British growers, for example, is collected from the south of France. Secondly, prior to sale to the growers, seeds are stored at low temperatures (usually below 10°C) under dry conditions. Thirdly, seeds committed for long-term storage in so-called 'gene banks' or 'germplasm banks' are stored at −20°C and at 5% moisture content. Under these conditions many commercially important seeds will retain their viability for very many years (see below). Fourthly, in stark contrast to the conditions in gene banks, the conditions used specifically for study of seed deterioration involve storage at 40°C and 20% moisture content. Under these conditions many seeds will lose viability within a year or less.

The conditions used in germplasm banks and those used to induce rapid seed deterioration have been selected following detailed analysis of loss of viability of seeds under various conditions. Equations are available for prediction of mean viability times at known temperatures and moisture contents:

$$\log \bar{p} = K_v - C_1.m - C_2.t$$

where $\bar{p}$ = mean viability period
m = moisture content as percentage of wet weight
t = temperature, °C
and K_v, C_1 and C_2 are constants which may be determined in the laboratory for a given batch of seed.

Further discussion of this is beyond the scope of this short volume, but any reader interested should consult either the excellent description of this topic given by Bewley and Black or some of the original work of Roberts (see suggestions for further reading). Some predictions of seed survival times, based on the equation, for seeds stored under 'gene bank' conditions are given in Table 5.2. For our purposes here, it is adequate to state that for each reduction in storage temperature of *c.* 5.5°C, the storage life of a given seed batch is doubled, as it is also for a decrease of 2% in the seed moisture.

Finally, in this section, it is necessary to consider survival of seeds under more natural conditions than gene banks. Few really long-term experiments have been carried out on survival under natural conditions, i.e. of seeds buried or partly buried in a moist state. The longest running experiment of this type was started in 1879 at the Michigan Agricultural College. The data from this experiment indicate that seeds of many wild species retain viability for 25 to 30 years and that some, such as seeds of various docks (*Rumex*) may retain viability for up to 80 years. The longest surviving species in the experiment was *Verbascum blattaria* (a close relative of the British species, *V. thapsus,* mullein), 20% of the seeds of which were viable after 90 years! It is clear then, that seeds of wild plants can survive ungerminated in the soil for many years. If this longevity is combined with seed dormancy and with the formation of seed banks (Chapter 4) then the species concerned will be very persistent and very difficult to eradicate as a weed from arable crops.

The latter point is emphasized by comparison with seeds of crop species. As indicated earlier, these have been subjected to selection against dormancy and for rapidity of germination. These features militate against longevity, and seeds of most crops, if maintained in a moist state, but in conditions otherwise unsuitable for germination, rapidly lose viability.

The data presented in this section clearly cast doubt on some of the claims made for germinability of long-buried seed, whether from an Egyptian royal tomb or from the wreck of the Mary Rose. In fact, most claims for germinability of seeds more than 100 years old cannot be substantiated either because the seeds did not in fact germinate or because the supposed age cannot be confirmed in any way. There are however, a few odd and therefore interesting instances of very old seeds being induced to germinate. Two such examples will be given. Firstly, some seeds of a leguminous tree, *Albizzia julibrissin,* stored in the herbarium of the British Museum since

Table 5.2 Predicted longevity of seeds under 'gene bank' conditions (−20°C, 5% relative humidity).

Species	*Time taken (years) for viability to fall to 95% of original value*
Pisum sativum (Pea)	1090
Vicia faba (Broad bean)	270
Lactuca sativa (Lettuce)	11
Allium cepa (Onion)	28
Triticum aestivum (Wheat)	78
Hordeum vulgare (Barley)	70

1793, germinated in 1940 when saturated with water whilst a fire (caused by an incendiary bomb) was being extinguished. Secondly, rattle necklaces found in tombs in Argentina were made of *Canna* seeds inside walnut shells. The *Canna* seeds must have been pushed into the walnuts whilst the nutshells were still soft. Then, as the nutshells hardened, the *Canna* seeds were trapped inside, forming rattles. The *Canna* seeds removed from one such necklace germinated and since the particular necklace was 600 years old (as judged from ^{14}C dating of the walnut shell), the strong implication is that these particular seeds had remained viable for 600 years. However, as already emphasized, such long-term survival of seeds appears to be extremely rare, and in the context of the commercial use of seeds it should again be noted that in the main, seeds of wild plants survive a good deal longer than seeds of cultivated plants.

5.2.2 Difficult or recalcitrant seeds

Although ideal storage conditions for most seeds are cool, dry conditions, some seeds rapidly lose viability in the dry state. This was brought home to the present author when he collected seeds of the creeping willow (*Salix repens*) for use in physiological experiments. The seeds were carefully maintained under 'ideal' conditions – cool and dry – for just a few days, after which they were set to germinate: none did! In fact, seeds of nearly all members of the genus *Salix* (willows) behave like this, as do seeds of other wild plants, including hazel (*Corylus avellana*) and horse-chestnut (*Aesculus hippocastanum*). Such seeds do, however, survive for up to several months in the moist state. The underlying physiological reason for this behaviour is not known, but it does have some relevance to forestry and agriculture. There is a small number of commercially important plants, including some willows (not just grown to supply the makers of cricket bats!), coffee and oil palm which have so-called recalcitrant seeds. Obviously season-to-season storage of these seeds is very difficult, and it is fortunate that these particular crops are relatively long-lived trees or shrubs which do not need to be planted for annual renewal. Even so, clonal propagation of such species either from cuttings, as in willow, or via tissue culture, as with oil palm, is clearly advantageous.

5.3 Modification of germination behaviour

Even amongst seeds of species used in agriculture and horticulture, where there has been extensive selection for lack of dormancy and for regularity of germination, there can be a good deal of variation in germination behaviour between, and even within, seed batches. The reasons for this are various and include factors such as differences in seed coat permeability, differences in residual dormancy (Chapter 1 and 2) and differences in vigour (Section 5.2.1).

What steps, then, can be taken to maximize the percentage, speed and regularity of germination? Some measures are obvious: seeds should be collected from plants grown in regions where seed can ripen and dehydrate properly; seeds should then be stored under suitable conditions. However, even when these practices are followed, the seeds of some species, mainly small-seeded 'cash' crops, still exhibit variability in germination, with resulting loss of profitability to the grower. Failure of seeds to germinate, for example, leads to gaps in the crop stand, whilst variability in the time taken to germinate leads to variability in the size and maturity of plants within the stand. Although this may seem trivial, it may actually be very important for batch-grown cash crops since harvesting the whole crop at one time relies on a relatively uniform development of the crop stand.

Although there are no widely applied answers to these problems, two relatively novel techniques are beginning to be used commercially. The first technique is to plant only seeds which will germinate within a given time and this can only be determined by first germinating the seed. Planting such germinated seeds by mechanical means presents problems not encountered in the mechanical sowing of dry seed; in particular, the emerged root (and the shoot, if germination proceeds further), are vulnerable to damage. To overcome these problems, the technique of 'fluid drilling' has been introduced. Seeds are germinated in water and then set into an agar gel. The agar gel may be gently extruded to give a linear array of germinated seeds, just as if the seeds had been planted normally in a row. Fluid drilling devices suitable for mounting on tractors have been developed both in Britain and in the USA; in Britain the technique is being used for some carrot and lettuce planting and in the USA for outdoor tomatoes. Manual fluid drilling techniques are also being used for some glasshouse crops.

The second novel approach relies on the fact that many of the physiological and biochemical changes which occur in imbibed seeds form a prelude to the resumption of active growth. These changes include both the repair and replacement processes, and the accumulation of enzymes and other macromolecules. Under certain conditions, many of the metabolic changes occur without resumption of growth. For example, seeds with a high temperature requirement for germination may imbibe water at temperatures too low for germination (i.e. resumption of growth) to occur. For tomato, this is just below 10°C. At this temperature, all the usual metabolic changes occur, as is evidenced by the accumulation of ribosomal RNA (Fig. 5-1), but there is no elongation of the radicle. If the seeds are now dried back, they are all in a state of readiness to resume growth rapidly. Re-imbibition of the seeds at a temperature suitable for germination leads to a higher percentage germination within a seed batch, a more rapid germination, as measured by a lower T_{50}, and a greater regularity of germination, as measured by the smaller spread of germination times (Table 5.3).

For seeds which do not have a high minimum temperature for germination, it is difficult to impose a temperature regime which allows a reasonable rate of metabolism whilst not permitting germination. For such

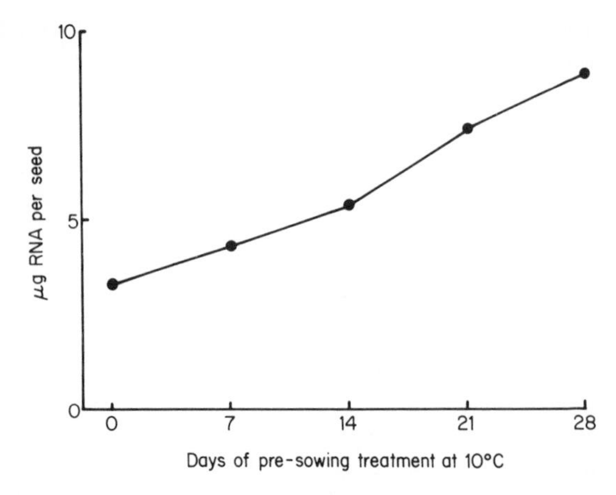

Fig. 5-1 Changes in RNA content of tomato seeds during low temperature pre-sowing treatment.

for example, celery, the pre-sowing treatment involves imbibition at a restricted water potential in solutions of metabolically inert osmotica such as polyethylene glycol. As with low temperature imbibition of tomato, this type of pre-sowing treatment prepares the embryo for growth. Drying back, followed by imbibition under normal conditions leads to the same types of improvement as described for tomato seeds.

These pre-sowing treatments have not reached the stage of development where they are suitable for routine treatment of large batches of seeds. Commercial application of the techniques is therefore at present somewhat limited, although pre-treated, dried back, pelleted seeds of tomato, pepper (*Capsicum*) and celery are commercially available in Great Britain.

5.4 Seeds as crops

5.4.1 Variety of usage

Man has always made extensive use of seeds in nutrition. The main crops used are firstly cereals, providing high-bulk, starchy staple foods. The seeds may be consumed more or less complete, although cooked, as is the usual practice with rice, or after processing to make flour (and hence bread or similar foods) as with wheat. Secondly, legumes or pulses (i.e. peas and beans) are also widely used. These two groups of plants still figure very largely in human nutrition, although in modern 'civilized' society it is all too easy to forget this. However, it is worth reminding ourselves that wheat flour, baked to make bread (and for some, pasta) is a regular component of our diet and that various other cereal seed products are also consumed regularly, for example as breakfast foods. Legumes such as peas, beans and 'baked beans' also feature in our diet. Furthermore, the ready availability of meat in the richer nations serves to blind us to the fact that hundreds

Table 5.3 Effect of low temperature pre-sowing treatment on germination of a median-vigour batch of tomato seeds at 17°C.

Seeds	*% germinated*	T_{50}*(h)	$T_{90} - T_{10}$†(h)
Control	91	120	55
Pre-treated	99	42	20

*T_{50} = time taken for 50% of the seeds to germinate
†$T_{90} - T_{10}$ = time taken from 10% germination to 90% germination

of millions of people in the world still rely very heavily on plants, and especially on plant seeds, as major components of their diet (Fig. 5-2).

In addition to this direct use of seeds or of milled seeds in nutrition, many different types of seed are used for more extensive processing to provide products for human and animal nutrition and also for industrial use. The range of seeds used in this way is too wide to provide an exhaustive list, but includes, for example, several oil-rich seeds, such as rape, soybean, oil palm and sunflower, used for the extraction of lipids to make margarine and cooking oils, or as lipid-rich additives for animal feed, or for industrial purposes such as the manufacture of lubricants, paints and soap. Protein-rich seeds may be processed to give purified plant protein; soybean is often used for this purpose, and in some countries where the climate is not suitable for soybean, such as Canada, peas are also used as a source of plant

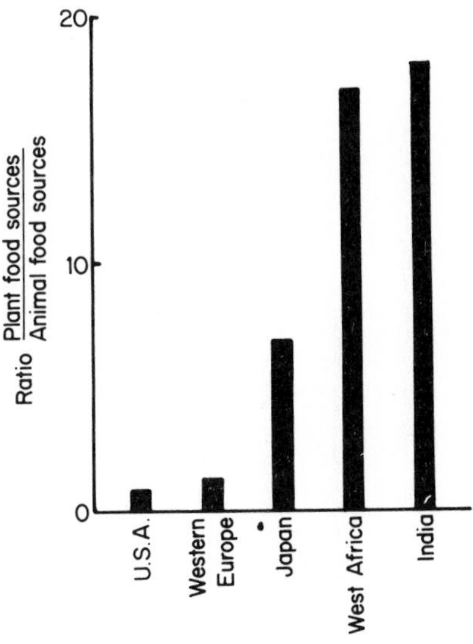

Fig. 5-2 Uses of plants and animals as food sources in typical diets of several different regions of the world.

protein. The protein is often flavoured and texturized to look like meat: apparently some people only find plant protein acceptable if it is disguised in this way! The storage polysaccharides may also be extracted to make packaged snack foods resembling crisps ('chips' in North America).

5.4.2 Nutritional and commercial value

The range of seeds used in human and animal nutrition, and for industrial purposes, is very wide, as is the range of specific uses. The term 'value' is therefore bound to have a variety of applications, some of the more important of which are dealt with here.

In the context of the use of seeds or processed seeds (such as flour or meal) for food, nutritional quality is all important. Three aspects of nutritional quality are considered here. Firstly, the carbohydrate:protein ratio of the food source is important. Carbohydrates are needed for the metabolic pathways which provide energy, and also to provide the basic carbon skeletons for various biosynthetic pathways. Protein is needed as a source of organically combined nitrogen for the biosynthesis of many important macromolecules, including nucleic acids and, of course, the body's own proteins. If the carbohydrate:protein ratio is too high, then too much bulk carbohydrate has to be ingested in order to meet the protein requirement. The need for protein in the diet varies, but as a general rule, children and adolescents (i.e. those who are actively growing) need more protein than adults. Figure 5-3 shows the carbohydrate:protein ratios for four seeds commonly used as food compared with those of fish and meat. It

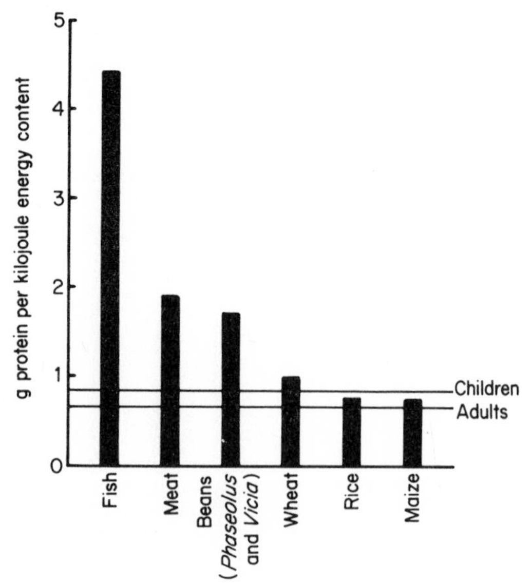

Fig. 5-3 Protein:energy ratios in various food sources. The average minimum ratios required for adults and children are shown by the two horizontal lines.

is again a salutary reminder that while the meat-rich diets of the more prosperous nations provide protein well in excess of needs, there are many people in the world whose dependence for food on a restricted number of plant sources makes their protein intake marginal.

The second factor to consider is protein quality. Proteins are synthesized from 22 different amino acids, and different proteins have these amino acids in different proportions. Plants, given a suitable source of nitrogen (such as nitrate, or an ammonium salt) plus a carbon source (usually generated in photosynthesis) are able to make all the necessary amino acids. Man is not so versatile. We can use ammonia as a nitrogen source (as in the re-cycling of ammonia from amino acid breakdown) but our major nitrogen needs are met by organically-combined nitrogen (particularly amino acids). Further, unlike plants, we cannot make all the amino acids. Thus, for man, there are certain amino acids which must be present in the diet. These essential amino acids are arginine, histidine, isoleucine, leucine, lysine, methionine, phenylalanine, threonine, tryptophan and valine. Unfortunately, the *bulk* seed proteins (i.e. the major storage proteins) of the major crops are all deficient in one or more of these essential amino acids, as illustrated in Tables 5.4 and 5.5. Of the major groups, cereal storage proteins are particularly poor in lysine, whilst legume storage proteins are poor in methionine. This does mean however, that in diets based largely or entirely on plants, a careful balance between cereals and legumes can provide an appropriate mixture of amino acids. Although our understanding of this is based on modern nutritional biochemistry, there is clear evidence from archaeology that more primitive societies empirically reached the conclusion that a balance between legumes and cereals was required. In pre-Christian times in the Middle East, chick peas (*Cicer arientinum*) were used alongside wheat and barley, whilst North and South

Table 5.4 Content of essential amino acids, plus cysteine, in storage proteins of pea (*Pisum sativum*) compared with typical animal protein. Data are moles per cent of total amino acid content.

	Storage proteins		
Amino acid	*Vicilin*	*Legumin*	*Animal protein*
Isoleucine	5.1	4.0	4.6
Leucine	9.2	8.1	8.3
Lysine	7.9	4.9	6.2
Methionine	0.22	0.65	1.9
Phenylalanine	6.2	4.9	5.4
Threonine	3.4	2.9	4.9
Tryptophan	—	1.1	2.1
Valine	4.6	4.6	5.8
Cysteine	0.35	0.71	1.1

Table 5.5 Comparison of essential amino acid content of various animal and plant food sources. Data are mg of amino acid per g nitrogen.

	Food source					
Amino acid	*Hen's egg*	*Beef*	*Fish*	*Maize*	*Rice*	*Soybean*
Isoleucine	393	301	299	230	238	284
Leucine	551	507	480	783	514	486
Lysine	436	556	569	167	237	399
Methionine	210	169	179	120	145	79
Phenylalanine	358	275	245	305	322	309
Threonine	320	287	286	225	244	241
Tryptophan	93	70	70	44	78	80
Valine	428	313	382	303	344	300

American Indians grew beans (*Phaseolus* spp.) alongside corn (i.e. maize).

The third example relating to nutritional quality concerns fatty acids; these are the long chain acids which occur in many different types of lipid, including the membrane lipids and the storage triglycerides. In saturated fatty acids the carbon valencies are totally satisfied throughout the chain, whereas in unsaturated fatty acids, double bonds occur in the hydrocarbon chain. Man is unable to introduce more than one double bond into a fatty acid chain, but needs, for membrane lipids, and for synthesis of more complex fatty acids, two particular fatty acids with more than one double bond. These fatty acids are linoleic acid (18 carbon atoms, 2 double bonds or 18:2) and linolenic acid (18:3). Plants are able to make both these fatty acids, and so the storage lipids of seeds are able to provide the dietary requirements for these components (Table 5.6).

In addition to nutritional quality, an evaluation of seed crops must include a consideration of the suitability of the crop for particular uses. Two examples illustrate this point clearly. The first example concerns wheat. In Britain, the major use of wheat flour is in the making of bread. However, wheat flour has a number of other uses, including the manufacture of pasta: spaghetti, macaroni, lasagna and so on. We think of wheat flour as being starchy, and rightly so, since it is indeed approximately 75% by weight starch. However, for the two different uses mentioned here, it is the protein component of the flour which is important. The cohesive properties of wheat dough, which are so important in providing the right texture, are due to the presence of a group of water-insoluble proteins known collectively as gluten, which comprises the major part of the wheat seed storage protein. The term gluten actually covers a very complex mixture of proteins consisting of two major classes, the glutenins (acid-soluble) and the prolamins (soluble in 70% ethanol). In wheat, the prolamins are also called gliadins. Both these classes contain several

Table 5.6 Examples of fat-storing seeds.

Species	*Fatty acid composition of storage triglycerides (moles %)*						*Uses*
	16:0	*18:0*	*18:1*	*18:2*	*18:3*	*Others*	
Soybean	12	4	25	52	7		Margarine; cooking and salad oils; ice-cream; paint; soap
Peanut	13	5	43	33		6	Margarine; cooking and salad oils; ice-cream
Sunflower	4	3	34	59			Margarine; cooking and salad oils; soaps; paints
Rape							
—low erucate	5	2	48	25	10	10[b]	Margarine; cooking and salad oils; cattle feed; lubricants
—normal	5	2	18	18		57[a]	
Cotton	22	2	31	45			Margarine; cooking and salad oils
Olive	6	4	83	7			Salad oils; preserving oils; paints
Linseed (flax)	5	4	10	43	38		Paints; varnishes
Castor bean	1		3	5		91[c]	Paints; lubricants; plastics; medicinal

a Erucic acid, 22:1. *b* Mainly 20:1. *c.* 18:1, OH, ricinoleic acid.

different (but relatively similar) individual types of protein molecule, and it appears that the cohesive properties of the proteins may be mainly ascribed to the glutenins. The wheats used for making pasta are the tetraploid *Triticum durum* group, in which the glutenins are not cohesive enough to make bread dough. In the hexaploid *Triticum aestivum*, bread wheat, the glutenins are more cohesive, and so provide the qualities needed for an elastic, extendible dough.

The second example concerns the storage lipids (triglycerides) of seeds, and more particularly their fatty acids. It has already been noted that the triglycerides of seeds are one source of the unsaturated fatty acids needed in human nutrition. However, the presence of considerable quantities of triglycerides in certain seeds has led to many other uses for such seeds (Table 5.6). In considering these uses, it is necessary to examine briefly the effect of desaturation (i.e. the presence of one or more double bonds) on the fatty acid molecule. The introduction of one double bond into the hydrocarbon chain causes the chain to be significantly kinked. This means in turn that unsaturated fatty acids do not pack together as well as saturated fatty acids. In other words, fatty acids with double bonds are more fluid than saturated fatty acids. Further, the fluidity increases as the number of double bonds increases, as is seen in melting points of the fatty acids shown in Table 5.7. In summary, plant lipids are in general more fluid than animal lipids. Thus, table margarines made from emulsified plant lipids retain their spreadability even when refrigerated, and cooking oils extracted from plants remain liquid, as compared with animal-derived cooking fats.

Table 5.7 Melting points of fatty acids.

Fatty acid	*Melting point* °C
16:0 Palmitic acid	63.1
18:0 Stearic acid	69.6
18:1 Oleic acid	13.4
18:2 Linoleic acid	– 5.0
18:3 Linolenic acid	– 11.0

However, factors other than fluidity are also relevant to the use of seed lipids. For example different fatty acid mixtures have different flavours. Thus, margarine made from sunflower seed lipids is readily distinguishable from margarine made from soybean lipids. The characteristic taste of butter arises partly from the presence of certain short and medium-chain fatty acids. So certain brands of margarine, made mainly from plant lipids, contain additions of these particular fatty acids in order that the margarine should taste more like butter. Further, the presence of significant amounts of particular fatty acids may make particular seed lipids unacceptable for nutritional purposes. The 18:3 fatty acids (i.e. fatty acids with 18 carbons and 3 double bonds) for example, are very prone to oxidation, leading to 'off-flavours'. Thus linseed (i.e. flax-seed) oil has a wide range of uses in industry (Table 5.6), but is never used for making cooking oil or margarine. Even the relatively small amount of 18:3 in soybean lipids means that antioxidants must be added to cooking oils and margarines made from this source.

In summary, seeds and seed products are used widely for nutritional and industrial purposes. Their usefulness lies in the range of storage compounds laid down during seed development, thus emphasizing the impact of seed physiology and biochemistry on the activities of mankind.

5.5 Crop improvement

5.5.1 Improvement in yield

One of the major aims of the grower is to maximize yield. This means in turn that one of the major aims in crop improvement is to increase the yield of a given crop species. Two general factors contribute to this, namely the genetic make-up of the plant and the environment (including the availability of nutrients) which interact to give the phenotype i.e. the plant as we see it. It is in the genetic factors that the plant breeder is mainly interested, and it is worth emphasizing here that plant breeding has been of primary importance in producing the modern high-yielding varieties of crops.

All crop plants used in modern agriculture and horticulture are the result of selective breeding, but certain cereal crops have been particularly intensively selected. Indeed, wheat and maize (corn) are probably the most intensely selected of the crops grown today. The hexaploid wheats in current usage are obviously much higher yielding than the earliest cultivated diploids (Fig. 5-4). The derivation of the hexaploid wheats is genetically complex, but the main point to emphasize is that the high ploidy level results in larger seeds (see Chapter 1), larger amounts of seed storage products and hence greater yields. An obvious ploy, then, is to use the hexaploids. This actually first happened about 2000 years ago, and thus provides an early example of selection.

However, the role of the plant breeder in producing high yielding lines of wheat (or indeed of any crop) is far more complex than simply selecting polyploids. The final yield of seeds in a flowering plant is a result of a very complex series of interactions, far from completely understood, between biochemical, physiological and developmental features, such as efficiency of photosynthesis, distribution of photosynthate, and cell number and size in the seed. Insofar as these are all at least partly under genetic control, they are all possible areas of selection by the breeder. The best example of this multifactorial approach to selective breeding is seen in the production of the so-called 'miracle' strains of cereals, high-yielding, disease-resistant plants with short stems (to prevent the weight of the 'ears' of seeds causing lodging or stem collapse). Thus for the breeder, seed physiology has to be considered alongside other aspects of the plant's physiology.

In passing, it should also be mentioned that in the end, nutritional deficiencies may defeat the objectives of the breeder. The availability of

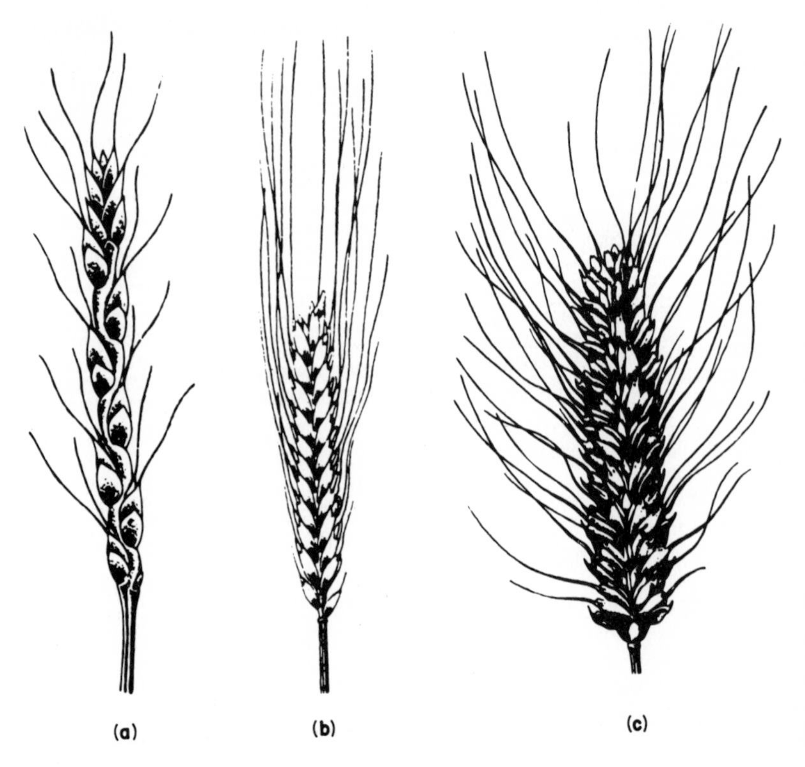

Fig. 5-4 Two 'primitive' diploid wheat species **(a)**, **(b)** compared with a modern hexaploid wheat **(c)**.

'miracle' strains of cereals did not, unfortunately, bring in the heralded 'green revolution' whereby the people of the world were to be fed. In order to produce large numbers of large seeds, cereal plants need applications of large amounts of fertilizers, particularly nitrogenous fertilizer. Even in the richer nations where fertilizers are used extensively, yields of some cereals (e.g. wheat in parts of Britain) could be improved by adding even more nitrogen. For poorer nations, it is sad, but true, that for the miracle strains to achieve anything like their true yield potential requires much more nitrogenous fertilizer than can be afforded.

5.5.2 *Improvement of quality*

As with yield, so the quality or value of the seed to the user is open to selection. The types of protein, carbohydrate and lipid laid down are genetically determined and again, therefore, are subject to mutation. In order to produce a crop with altered quality or value, the breeder needs lines or strains carrying mutations in respect of seed storage products. Selective breeding for quality has certainly been successful where a single gene controls deposition of a particular product. A good example of this is seen in oil-seed rape. Up to about 20 years ago, very little rape was grown in Europe, and what was grown was not used much in human or animal nutrition, but rather in lubricants such as cutting oils. The seed lipids were characterized by a particular fatty acid, erucic acid (22:1) which was thought to be harmful to mammals if ingested in any quantity. The ability of the seed to make this lies in the ability to elongate 18:1 by successive additions of two 2 carbon (2-C) units. A mutation in the gene coding for the enzyme involved, to prevent this addition of 2-C units, will obviously prevent the accumulation of erucic acid. Breeders have made use of this, and the availability of 'zero-erucate' rape-seed has been followed by a huge expansion in its use as a crop in Western Europe, as is evidenced by the brilliant yellow fields in May, June and early July, when the plants are flowering. As is shown in Table 5.6, the zero-erucate varieties lay down large amounts of 18:1 and 18:2 (and some 18:3) fatty acids which are perfectly acceptable for mammalian nutrition (with the proviso that steps have to be taken to prevent oxidation of 18:3 acids in margarine and cooking oils: see Section 5.4.2). However, this plant breeding success story must be tempered with some reservation: nutritional biochemists today are rather less certain than they were 25 years ago that erucic acid is at all harmful to man or to his farm livestock!

Of more general importance than rape seed oil to human nutrition are of course the seed proteins of the cereals and legumes. The deficiency of these seed proteins in certain essential amino acids has already been noted. It is then another area where selective breeding can have an impact? Unfortunately, the answer here is rather pessimistic. In dicotyledonous plants, such as legumes, the seed storage proteins are globulins, which are proteins soluble in salt solution or in acid. Seed globulins in dicotyledons fall into three general classes according to size: the three classes have sedimentation coefficients (*s*) (measured in the ultra-centrifuge) of 11–12*s*, 7–8*s* and 2*s*, and although the names given to these proteins differ between species, the equivalence of these classes right across the Dicotyledones is readily seen. Within a given species, each class exhibits some heterogeneity, and this is at least partly due to the fact that there are several similar genes coding for each class of protein. For example, there may be about five different (but very similar) genes each coding for a 7*s* storage protein molecule, thus giving rise to a 7*s* storage protein mixture of five different (but nevertheless very similar) individual types of protein molecule. Overall, in a typical legume, such as pea, the storage globulins may be encoded in ten to twenty different genes. To make a significant difference to the methionine content of the globulin fraction, for example, would therefore require a similar mutation to occur simultaneously in several different genes. The absence of suitable lines for selection by the breeder is evidence for the improbability of such an event.

The storage protein complement of cereal seeds is even more complex than that of legumes and other dicotyledons. In cereals, the salt- or acid-soluble globulins make up about 20–30% of the storage proteins, whilst the glutenins (another group of acid soluble proteins) and the prolamins (alcohol-soluble proteins) together make up 70–80% of the seed storage proteins. The low lysine content of the cereal storage proteins is ascribable almost entirely to the glutenins and prolamins. As with the storage proteins of dicotyledons, these proteins are actually encoded in small families of genes, each gene in the family coding for a particular member of its particular class of proteins. Since both the glutenins and the prolamins are complex classes of protein in that they are made up of many different, but similar types of protein molecule, it is quite possible that 20 to 40 genes are involved. Mutation in a single one of these genes would not make a significant difference to the lysine content of cereal seed protein. However, single gene changes are known in several cereal species which apparently increase the lysine content of the seed proteins. These 'high-lysine' mutants are characterized by low ratios of glutenin + prolamin:globulin. In other words, the apparently high lysine content is caused by lower than usual amounts of the major storage proteins. Unfortunately, this is not balanced by an increased globulin content, so the seed protein content is actually lower than usual in these varieties. The mutation which causes this presumably occurs in a single gene which regulates the deposition of the glutenin and prolamin fractions. In some of these mutant lines, the starch content of the seed is also lower, suggesting a mutation in a gene having a major effect on seed development.

5.5.3 Prospects for the future

The role played by 'traditional' plant breeding in the provision of suitable strains or lines of crop species for modern agriculture must not be underplayed. However, the desirability of radically improving the nutritional quality of seed storage proteins, or of changing the composition of seed lipids for particular industrial or nutritional uses, has led to the development of other approaches to crop improvement. The transfer of 'foreign' genes into plant cells by genetic engineering techniques is now a reality, as is the growth of whole plants from such genetically engineered cells. This has led to ideas that an insertion of 'foreign' genes may be a quicker route to the improvement of seed quality than the more usual plant breeding methods. For example, if a copy, or better, several copies, of the egg albumin gene (Table 5.5) could be inserted into a cereal or legume plant, the plant would have the genetic potential to make that protein. However, there is a great difference between putting new genes into a plant chromosome and getting those genes to work at the right time. In order for the inserted genes to be useful, they must be active during seed development. Further, they must be very much more actively transcribed than the genes coding for the enzymes and other 'working' proteins (as described in Chapter 1). In other words, an inserted foreign gene coding for an ideal protein must behave as a normal seed storage protein gene. We are still a long way from understanding how gene activity is regulated in eukaryotic organisms. However, there is clear evidence for some genes that DNA sequences adjacent to the gene itself (and particularly those sequences 'upstream' of the gene i.e. before the coding sequence) are involved in regulation. This has led to the successful insertion of a gene coding for the bean (*Phaseolus*) storage protein, phaseolin, together with the gene's 'upstream' regulatory sequence into tobacco plants. In this experiment the presence of the regulatory sequence caused the gene to be expressed at a high level only in the tobacco seeds, whereas without the regulatory sequence, the gene was expressed at a low level in all parts of the genetically engineered tobacco plants. Tobacco is selected at present for such experiments because it is particularly easy to regenerate whole plants from genetically engineered cultured cells. Experiments of this type have not at the time of writing, been successfully carried out with any of the important food crop plants. However, there is a clear indication that manipulation of seed nutritional quality, and other seed characters, by genetic engineering, is an achievable goal.

Further Reading

BEWLEY, J. D. and BLACK, M. (1978). *Physiology and Biochemistry of Seeds, Vol. I.* Springer–Verlag, Berlin and New York.

BEWLEY, J. D. and BLACK, M. (1982). *Physiology and Biochemistry of Seeds, Vol. II.* Springer–Verlag, Berlin and New York.

BRYANT, J. A. (1981). Differentiation of storage cells in seeds of legumes, in *Quality in Stored and Processed Vegetables and Fruit,* eds P.W. Goodenough and R.K. Atkin, pp. 193–202. Academic Press, London.

BURGASS, R. W. and POWELL, A. A. (1984). Evidence for repair processes in the invigoration of seeds by hydration. *Annals of Botany,* **53**, 753–7.

DURE, L. S. (1975). Seed formation. *Annual Review of Plant Physiology,* **26**, 259–78.

GRAY, D. (1981). Fluid drilling and vegetable seeds. *Horticultural Reviews,* **3**, 1–27.

GRIME, J. P. (1975). *Plant Strategies and Vegetation Processes.* Wiley, Chichester.

GRIME, J. P., MASON, G., CURTIS, A. V., RODMAN, J., BAND, S. R., MOWFORTH, M. A. G., NEAL, A. M. and SHAW, S. (1981). A comparative study of germination characteristics in a local flora. *Journal of Ecology,* **69**, 1017–59.

HEISER, C. B. (1973). *Seed to Civilization.* Freeman, San Francisco.

HEYDECKER, W. and COOLBEAR, P. (1977). Seed treatments for improved performance – survey and attempted prognosis. *Seed Science and Technology,* **3**, 881–8.

KENDRICK, R. E. and FRANKLAND, B. (1982). *Phytochrome and Plant Growth, 2nd edition.* Edward Arnold, London.

LAIDMAN, D. and WYN JONES, G. eds (1979). *Recent Advances in the Biochemistry of Cereals.* Academic Press, London and New York.

MILLERD, A. (1975). Biochemistry of legume seed proteins. *Annual Review of Plant Physiology,* **26**, 53–72.

OPEN UNIVERSITY S202 COURSE TEAM (1981). *Plant Physiology II.* Open University Press, Milton Keynes.

ROBERTS, E. H. (1973). Predicting the storage life of seeds. *Seed Science and Technology,* **1**, 499–514.

SLATER, R. J. and BRYANT, J. A. (1982). RNA metabolism during breakage of seed dormancy by low temperature treatment of fruits of *Acer platanoides* (Norway maple). *Annals of Botany,* **50**, 141–9.

SUSSEX, I. M., DALE, R. M. K. and CROUCH, M. L. (1980). Developmental regulation of storage protein synthesis in seeds, in *Genome Organiza-*

tion and Expression in Plants, ed. Leaver C.J., pp. 283–90. Plenum, New York and London.

SUTCLIFFE, J. F. and BRYANT, J. A. (1977). Germination and seedling growth, in *The Physiology of the Garden Pea,* eds. Sutcliffe J.F. and Pate J.S., pp. 45–82. Academic Press, London and New York.

THOMPSON, J. R. (1979). *An Introduction to Seed Technology.* L. Hill, Glasgow and London.

TILLBERG, E. and PINFIELD, N. J. (1982). Changes in abscisic acid levels during after–ripening and germination of *Acer platanoides,* L. seeds. *New Phytologist,* **92**, 167–72.

VILLIERS, T. A. (1975). *Dormancy and the Survival of Plants.* Edward Arnold, London.

Index